BUILD-IT-YOURSELF

Solar Air Heater

Mounts on Any Sunny Wall, Turning Solar Energy into Free Heat for Your Home

RODALE PLANS

by
Ray Wolf

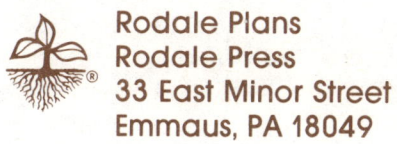
Rodale Plans
Rodale Press
33 East Minor Street
Emmaus, PA 18049

Copyright © 1981 Rodale Press, Inc. All rights reserved. No part of this publication may be reproduced or transmitted in any form or by any means, electronic or mechanical, including photocopy, recording, or any information storage and retrieval system, without the written permission of the publisher.

Rodale Plans
Rodale Press
33 East Minor Street
Emmaus, PA 18049

Printed in the United States on recycled paper containing a high percentage of de-inked fiber.

Library of Congress Cataloging in Publication Data

Wolf, Ray.
   Solar air heater.

   (Rodale plans)
   1. Solar space heating. I. Title. II. Series
TH7413.W64   697'.78   81-5860
ISBN 0-87857-359-3 paperback   AACR2
2 4 6 8 10 9 7 5 3   paperback

Projects and plans designed and published by Rodale Press have been researched and tested by its in-house facilities. However, due to the variability of all local conditions, construction materials, and personal skills, etc., Rodale Press assumes no responsibility for any injuries suffered or damages or other losses incurred during or as a result of the construction of these designs. We recommend that before major construction is undertaken the plans are first reviewed and approved by a knowledgeable, local architect or builder for feasibility and safety as well as compliance with all local and other legal and code requirements. All instructions and plans should be carefully studied and clearly understood before beginning any construction.

Technical Illustrator
Frank Rohrbach

Project Designers
Dennis Kline
LaMar Laubach
Frank Rohrbach

Thermal Engineer
Robert Flower

Product Testing
David Sellers
Harry E. Wohlbach

Copy Editor
Felicia D. Knerr

Book Layout
Merole Berger

Photographers
John Hamel
Ray Wolf

Table of Contents

SECTION I
Introduction 5
Chapter 1 Design and Operation 7

SECTION II
Introduction 13
Tools 14
Materials 15
Chapter 2 Building the Collector Box 19
Chapter 3 Building the Glazing Frame 35
Chapter 4 Installing the Heater 53

SECTION III
Blueprints 64

Section I Introduction

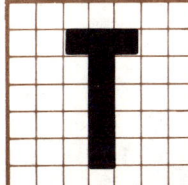

This book is your ticket to solar energy. It is one of the best ways I know for you to save both energy and money in the coming years. We've researched the topic and are confident this is the most cost-effective and best build-it-yourself solar collector available.

We think the concept of a plans book will ensure that you have a successful solar experience. This plans book truly has been a group effort. From my first idea for the book, a lot of people have worked a lot of hours to make it a reality. Frank Rohrbach not only did all the fine illustrations, he also worked on his own time to digest the existing literature and come up with the basic design of the collector. In many ways, this book is more his than mine. LaMar Laubach built our prototype and, to prove it worked, installed it on the front of his house. During the dead of winter, he came in every day with a warm smile on his face telling us all how well "his" unit had performed the day before.

From that early design, our Director of Design, Jim Eldon, worked with Dennis Kline to perfect the unit. Our Thermal Engineer, Robert Flower, constantly offered his advice, and the final design was improved a lot, thanks to his expertise. David Sellers was chiefly responsible for testing the unit. He worked with Bob to develop the information needed to modify and perfect the design.

We installed four units on our building, and David and Bob began the long process of monitoring their performances. Each unit had minor design differences. Some ideas were quickly rejected, others were adopted, and still others were improved upon. After more than six months of testing and monitoring, we knew we had the design we wanted.

The collector presented here uses only commonly available materials, is easily constructed, performs better than all but the most expensive of commercial units, and should easily last 20 years with minimal upkeep.

We feel this is somewhat of a breakthrough in the solar field. No one we know of has put this much time and effort into a unit designed to be owner-built. We at Rodale feel owner-built is the best choice for solar energy units, and this book is our effort to help you solarize your life and home.

This collector can be built easily and will work. Follow the instructions, blueprints, illustrations, and photos, and you should have no problems. I've built and installed the units I've written about, and Frank Rohrbach has designed and built the subject of his fine drawings.

I owe thanks to Kim Greenawalt, as well as to all the people I've already mentioned. Kim not only typed the manuscript, she also kept things running while I was busy writing or out installing hot air heaters.

To everyone who helped make this book a reality, I say thank you. To you who have bought the book, I wish you well in your coming adventure. You have bought what I sincerely believe to be a good book that explains a very well designed solar collector. If you follow our advice, you'll end up with a working solar collector for a fraction of the cost of a commercial unit.

May you have many warm and happy solar-heated days after you've built your collector. May you enjoy building your collector as much as I have enjoyed working with the people who made this book possible.

Ray Wolf

1 Design and Operation

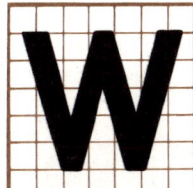

What is solar-heated air, and why would anyone want to collect it and spread it around their house?

Solar-heated air is the air inside a parked car on a sunny summer day. Your car acts as a solar hot air heater by accident. Conceivably, you could park your car next to your house and simply run a pipe through a window to bring in solar-heated air to help heat your house, but we think we've got a better idea. We've designed an attractive collector that effectively collects and distributes heat around your house.

Our collector is designed to mount directly on a sunny wall of your house, with through-the-wall vents to bring cool room air into the collector, heat it, and then return it to the room. The goal is to grab every possible Btu (a unit of thermal energy) and spread them around your living area to replace Btu's generated by other, more expensive, sources of energy, such as gas, oil, electricity, or wood.

The collector works on the same principle that your car does when it overheats on a sunny day. This principle is known as the greenhouse effect. Illustration 1-1 shows how the collector works. Sun shines through the glazing and hits the collector surface. We use a fiberglass-reinforced plastic glazing and two layers of common aluminum window screen, painted black, as the collector surface. Your car uses its windows as the glazing and the interior as the collector surface. In your car, the seats and dashboard absorb the sunshine and, in turn, heat the air and you. In our collector the two layers of screen trap about half the sunshine that enters the collector, while the rest passes through the screen and is absorbed by the black back of the collector box. These two collector surfaces then give off their heat to air circulating through the collector. In your car, air is trapped inside and becomes superheated. In the collector, room air is pulled in at the bottom of the collector, is warmed as it is pulled through the collector by fans, and is returned to the room as heated air. The principles are the same, but the collector is much more efficient than your car.

Remember, sunshine does not warm air directly; warm surfaces do. That is why, if you stand outside in the sun on a clear winter day, you will be warm even though the surrounding air is cold. The collector uses this principle to the maximum to produce as much heat as possible from the available sunshine.

Just as your car's interior doesn't get as hot in the winter as in the summer, the collector does not get as hot on cloudy days as on sunny days. We should say right off that the collector is not a total panacea for your heating bills. It only works when the sun is shining. At night it does nothing except serve as added insulation to your wall; it produces no heat when the sun is not up.

The collector's superefficient design enables it to produce lots of energy, even on cloudy days. In fact, the performance of the unit on what you might consider a cloudy day will be quite a surprise. You do not need blinding sunshine to produce heat.

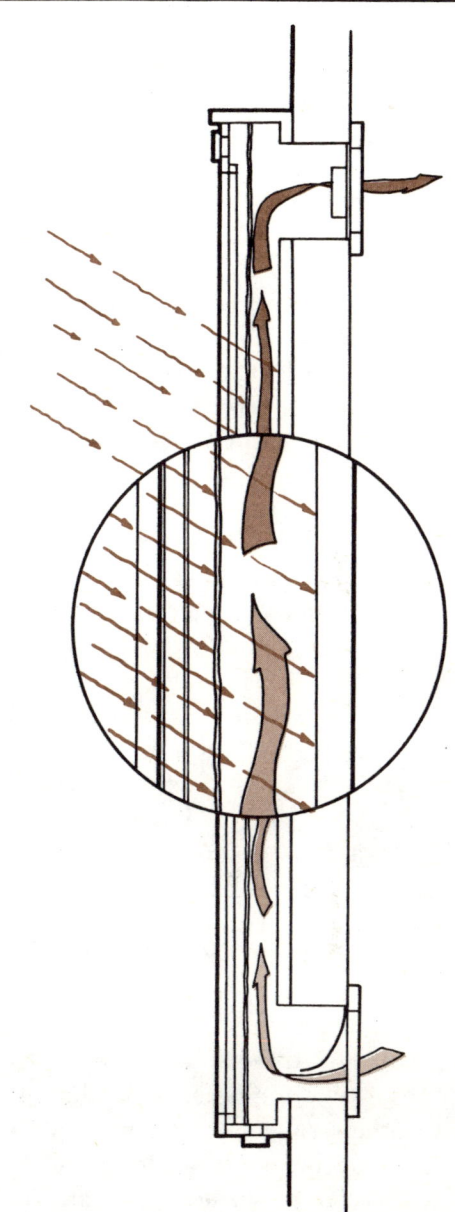

Illustration 1-1—The solar collector traps the sun's energy and uses it to heat air, warming the house.

When talking about the efficiency of a solar collector, you have to consider not only how much energy you collect, but also how much it costs you to collect it.

Let's look at two parked cars, each collecting solar energy in a parking lot on a sunny summer car. Car A is a used VW costing $2,500, while car B is a "previously owned" deluxe Rolls Royce costing $42,000. Let's say our VW collects the solar equivalent of 1 gallon of gasoline, while the Rolls collects the equivalent of 1½ gallons of gasoline during the day. Which is the better deal? The cost per gallon of the VW gasoline is a fraction of that of the Rolls; thus, although the cheaper collector doesn't collect as much energy, the cost per gallon is far below the more efficient Rolls Royce collector.

We feel we've designed a collector that works like a Rolls Royce at a VW price. Our collector is the most cost-effective solar collector we know of. If you build your unit with entirely new materials it should cost you no more than $250, installed. The cheapest comparable-size commercial unit sells for slightly over $400, delivered but not installed. Both units qualify for a 40 percent federal tax rebate.

When you talk to solar salesmen, they will talk one of two numbers, depending on which favors their collector: cost per square foot of collector surface or Btu's delivered per square foot of collector surface per year. The first is an indicator of the cost of the unit, the second, an indicator of the overall effectiveness of the unit. A combined figure gives you the cost-effectiveness of the unit—sort of a

Annual Fuel Savings

ZONE	FUEL OIL (gallons)	NATURAL GAS (therms)	COAL (tons)	ELECTRICITY (kilowatt-hours)	WOOD (cords)
1	25	29.7	0.13	610	0.26
2	30	35.6	0.16	731	0.31
3	35	41.6	0.19	853	0.36
4	40	47.5	0.21	975	0.42
5	45	53.5	0.24	1,097	0.47
6	50	59.4	0.27	1,219	0.52

Chart 1-1—Annual fuel savings from Rodale's solar hot air heater in the active mode. Use illustration 1-2 to find your zone, then look across to see how much of your type of fuel your solar energy is worth. Multiply the amount of fuel by its cost to get your annual savings in dollars.

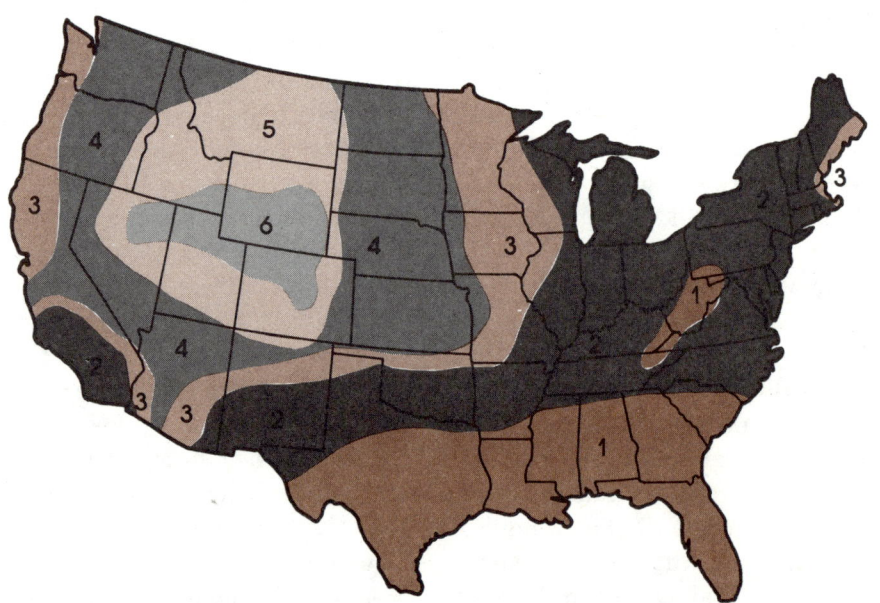

Illustration 1-2—Solar energy zones of the United States. Find which zone you are in, then use chart 1-1 to calculate your potential annual fuel savings from the collector.

solar "MPG" rating. Very few salesmen want to talk these numbers.

For our collector, with the fans installed, the numbers look like this. (These figures are based on a $150.00 installed cost for the unit—$250.00, minus the 40 percent federal tax rebate—for our central Pennsylvania climate with 5,800 degree-days in an average year. With 29 square feet of glazing, we have a cost per square foot of $5.17. The collector's energy output in this climate is about 80,000 Btu's per square foot, per year. If we make the very conservative assumption that the collector has a 10-year service life, the 29-square-foot unit will produce about 23.2 million Btu's during its lifetime. This represents an energy cost of only $6.46 per million Btu's, guaranteed *free of inflation* over the assumed 10-year life of the unit. That is equal to fuel at 60 cents per gallon, wood at $59.00 per cord, and anthracite coal at $115.00 per ton. That is a highly efficient, highly cost-effective collector, sort of like a car that has an 80 MPG city rating. No one can match those numbers, and the advantage is even greater if you use our actual estimated 15- to 20-year life expectancy.

Chart 1-1, used with illustration 1-2, shows the annual fuel savings generated by our collector for all areas of the country. Commercial units simply cannot compete when you consider cost of labor and of shipping materials all over the country. By using your own labor and by building your collector where it will be used, you will have the most cost-effective solar collector possible.

What's so special about our design? The collector is sized to use all dimensional lumber. The overall size of the collector is 4 feet by 8 feet, the same as a full sheet of plywood. The sides and all other parts are made with standard-size lumber. This is not only cheaper but makes the unit much easier to build as well.

We've designed the unit so it mounts

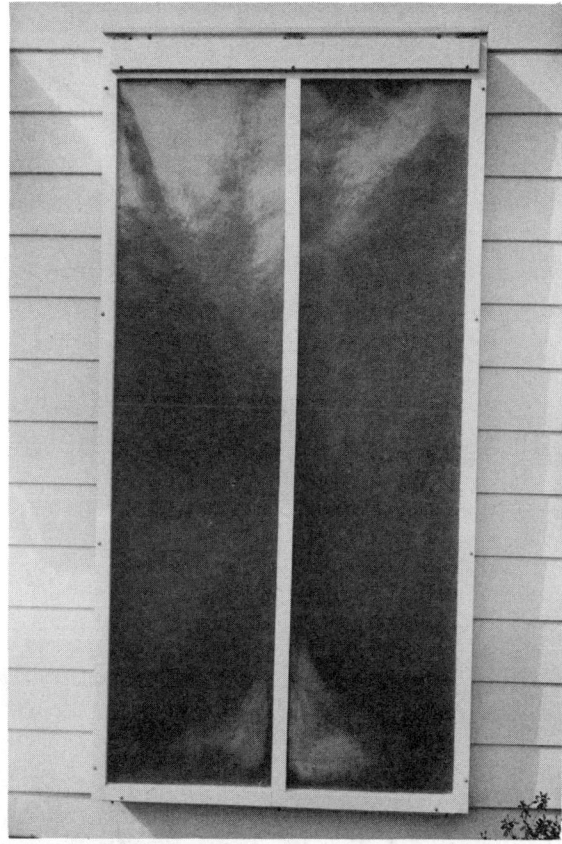

Photo 1-1—A vertical installation of Rodale's solar hot air heater on a frame house with aluminum siding.

either vertically or horizontally, as shown in photos 1-1 and 1-2. The two units are almost identical, with only slight variations. We've used a modular concept to make the collector adaptable to more locations.

If you leave the windows of your car open slightly during the summer, the inside of the car doesn't get as hot. We take advantage of this principle by constantly pulling room air into the collector with a fan. That way, the air exiting the collector is warmer than room air, but not greatly so. Our goal is to have the air leaving the collector heated to between 80° and 95°F. If you heat the air hotter than that, you will have two problems. First, a lot of the energy you collect will radiate back out of the collector before it gets into the house, and second, the superheated air you do move into the room will cause an uncomfortable hot spot.

We've monitored a completely closed

Photo 1-2—A horizontal installation of Rodale's solar hot air heater on a masonry home with stucco.

collector on a December day. With moderate sunshine and an ambient outside temperature in the thirties, the inside of the collector will reach 200°F before noon. Temperatures this high are a waste of energy. With the thermostatically controlled fan system we've designed for the collector, the average temperature of the collector surface should be slightly over 100°F, and the exiting air should be a comfortable 85°F. The lower air temperatures actually enable a lot more energy to be collected. It's the old "a lot of a little adds up to more than a little of a lot" principle.

A lot of similar-style collectors do not use fans. The feeling is that a totally passive unit is better than one that uses electricity to operate. We have found that passive solar designs are indeed best for new homes or major reconstruction. But to be truly effective, passive designs need some type of thermal mass in which to store their peak heat output for later use. This runs into expensive storage problems.

In testing, we found that using two fans almost doubles the overall efficiency of the collector, compared to a passive collector. This greatly improves the overall cost-effectiveness of the unit. The two fans together require only 0.048 kilowatts of electricity per hour. If the fans run an average of seven hours a day for five months a year, they will cost only $3 per year at an electricity cost of six cents per kilowatt-hour—a bargain price for the amount of heat you will receive. Plus the fans will increase the life expectancy of the unit by reducing the temperatures within the collector and will make the room you heat with the collector much more comfortable.

Many other units use only one layer of glazing, to reduce the cost of the unit. We think this is short-sighted economics. Illustration 1-3 indicates which area of the country needs one layer of glazing and which area requires two layers. If in doubt, use two layers. If you will be using your unit without fans, the temperatures inside the collector and the potential heat loss will be greater—thus the need for an added layer of glazing in swing areas.

We also found that making your own vent grilles saves money and adds 10 to 20 percent efficiency to the collector. Commercially available grilles restrict airflow too much, building up and wasting heat in the collector.

We've built in a summer venting system to help extend the life of the collector. Without such a venting system, summer temperatures will build up to well over 250°F within the collector. Illustration 1-4 shows the summer mode for the collector. All it requires is that, after the heating season, you open the two vents and unplug the fans. Next heating season, bolt the vents shut, plug in the fans, and you are ready to go. Without the vents, the excessive temperatures would damage not only the wood but the fans as well.

Illustration 1-3—Glazing zones of the United States. Zone 1 needs only single glazing, while zone 2 needs two layers of glazing. The unnumbered zone is a swing area. Collectors in this area need two layers if used without fans (passive), one layer if used with fans.

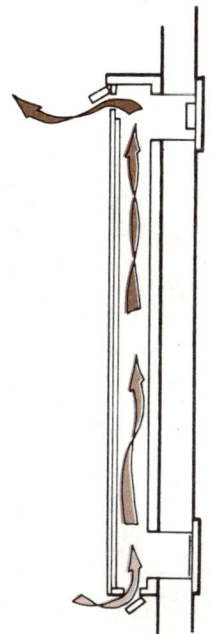

Illustration 1-4—The two vent doors should be opened in the summer to maintain a constant airflow through the heater. This prevents the unit from overheating and from heating the house.

ORIENTATION

Due to the flat design of the collector, it can be mounted on almost any wall. If there are no obstructions, and if you are willing to accept some decreases in efficiency from orientations that are not directly south, every house has a properly oriented wall.

To find south for your home, first calculate solar noon. Find out sunrise and sunset times for a particular day. Solar noon occurs exactly halfway between these times. At solar noon, drive a stake in the ground, and then drive a second stake at the end of the shadow of the first stake. The shadow will point exactly south. (This will differ from magnetic south.) Then you can calculate the loss in efficiency from chart 1-2, based on how much off of solar south your wall faces.

Estimate the potential shadows on your site at noon during the winter, to see if you have any major obstructions. (Summer obstructions are unimportant, as you won't want to produce heat during the summer.) Winter is the only time you are concerned with.

OPERATION

To operate your collector, all you have to do is install it and let it go. If you build it exactly as we detail, the thermostat will turn the fans on when it is producing heat and off when it is not.

Over time, dust and other materials will build up inside the collector, reducing its efficiency. Every couple of years, or more often if you notice a problem, you should remove the glazing cover and vacuum out the inside of the collector and the collector screens and clean the glazing. Some glazing materials need to be treated every couple of years to prevent deterioration; follow the manufacturer's suggestions. The unit will need occasional repainting; the weather stripping will need to be replaced, as will the back-draft damper. The plastic flappers of the damper

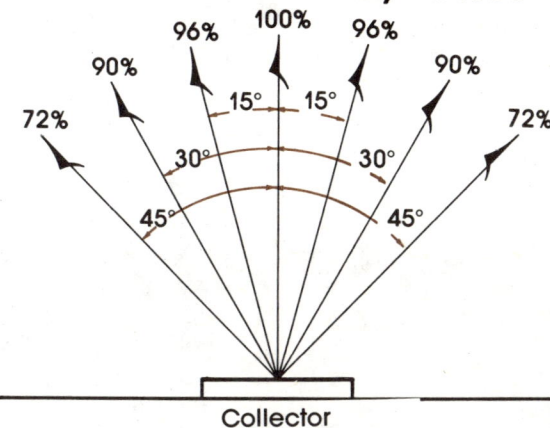

Chart 1-2—Collector efficiency losses due to less-than-ideal orientation. If your collector is exactly perpendicular to solar south, you have no loss. If your wall is off, estimate by how many degrees, to see how much it will hurt your collector's performance.

will degrade every couple of years, so check on them occasionally. If you feel air seeping into your room around the vent at night, replace the flappers.

That is it. The unit is designed to be as carefree as possible. We don't want you to become a slave to your collector, and most likely you don't wish such a fate either. We've designed a very efficient, easy-to-build collector that we are sure you will be satisfied with.

We wish you many warm and comfortable years of solar heating.

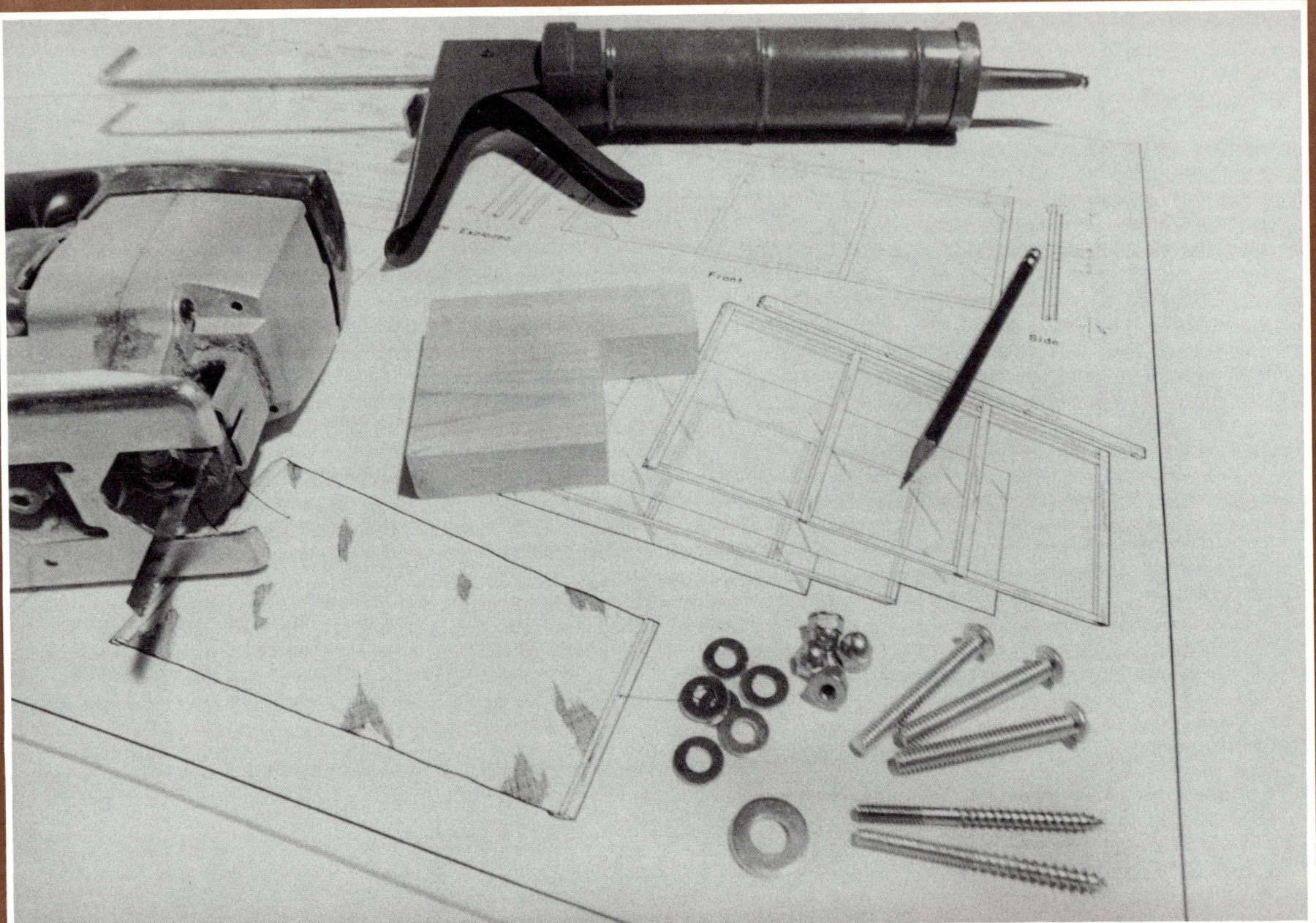

Section II Introduction

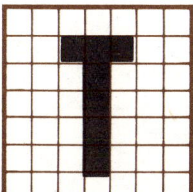

This is the hard part of the book. This is where you not only read but you work as well. However, this part of the book pays big dividends for your work. You'll end up with a solar hot air heater installed on your house, lower fuel bills, and a warm feeling—not only in your home but also within yourself—when you are done.

Before you start cutting up every board in sight, let's go over the layout of the book, so you understand what you are about to work with. Part I described the collector and how to operate and maintain it. Following this introduction you'll find the Tools and Materials listings. The Tools list is designed to let you know what tools you will need to complete the collector. We have not included the tools needed for installing the collector. These will vary greatly, depending on your home's construction.

Likewise, the Materials listing does not include items for installation. Your home's construction will dictate the materials you need. Information on the type of fans you will need, if you decide to use them, is on blueprint sheet 5.

The rest of this part of the book is in three chapters; two of them explain the building of the collector, the third explains the installation of the unit. Each chapter is designed to be used in conjunction with the blueprint sheets at the end of the book.

Chapter 2 details how to build the collector box, which is the back part of the finished collector. The first part of the chapter explains construction of a vertical collector; the second, a horizontal collector. The collector boxes are very similar and use identical construction techniques, but there are differences; pay attention to which type of collector you are building, and ignore the other instructions. The vertical collector box is detailed on blueprint sheet 1; the horizontal collector box is on blueprint sheet 2. Blueprint sheet 6 has the cutting diagrams for all the lumber used for both types of heaters.

Chapter 3 explains building the glazing frames for both vertical and horizontal heaters. The glazing frames are wooden frames with two layers of fiberglass-reinforced plastic attached to them. They allow sunlight into the collector but prevent cold air from entering, thus enabling it to produce heat. The first part of the chapter explains a vertical glazing frame and should be used with blueprint sheet 3, while the second part of the chapter explains a horizontal glazing frame and should be used with blueprint sheet 4.

Chapter 4 is devoted to the installation of your collector. This chapter is different from the others; it does not give specific instructions. Due to the many variables of home construction, we present a sequence of construction and guidelines but no specific step-by-step directions. We cover two types of home construction: frame walls and masonry walls. Read over the chapter first; if you don't think you will be comfortable trying to install the unit yourself, ask around for help, or hire someone to work with you. It should only take about three hours to install a unit on most walls.

The last part of chapter 4 covers building duct work to carry the heat into your home and connecting the fans and back-draft damper. Every collector must have a back-draft damper, but the fans are optional. Chapter 1 will help you decide whether or not you want fans.

That completes the book. Work carefully, and have a helper where noted. Your main helper will be this book. Everything you need to know is spelled out in the coming pages. Read carefully, keep the book on your workbench, and use the blueprint sheets, and you should have a successful solar heater in no time. Now, get ready to build your heater. Go through the Tools and Materials sections and the lumber list on blueprint sheet 6 before shopping, then get your materials and start building your solar hot air heater.

Materials

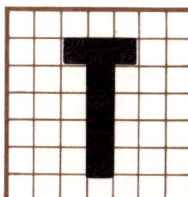
This is the expensive part of the book. Get out your checkbook and get ready to buy your materials. You should be able to buy everything except the glazing material at your local lumberyard and hardware store. Most likely you'll either have to mail-order the glazing materials or contact the manufacturer for the name of a local distributor. If you will be using fans with your unit, you may have trouble locating the proper type of fans and thermostat. We give the specifications for these, as well as the names and addresses of the fan and thermostat manufacturer, on blueprint sheet 5. Contact them for a local distributor.

The materials listed here include everything you will need for building your collector, but not the materials you will need for installing it. The materials required for installation will vary greatly, depending on the construction of your home. See chapter 4 for an idea of what you will need for installing the collector.

You'll notice a series of colored blocks throughout this section. They are for you to record your expenses. Keep track of every penny you spend on your solar collector, including installation costs. It is all eligible for the Federal Energy Tax Credit. At this writing, the credit is a full 40 percent. Keep a complete ledger of your expenses, and for every dollar you spend, you'll get back 40 cents at tax time. You'll notice we've included the cost of this book. As a plans book, with blueprints, the cost of the book qualifies as a valid part of the cost of the collector and is eligible for the tax credit if you build the collector. Be sure to check with your state energy office for details on any state tax credits. Your own labor is not eligible, but any labor you pay for is. Keep all your receipts in the blueprint envelope at the back of the book.

Go through each materials entry and become familiar with what you will need, then go shopping. To save on energy, you may want to phone area suppliers before leaving home.

Note that the vertical and horizontal collectors need different quantities of materials in some cases. Where the quantities are the same, there is just one listing; where they differ, there are two.

Lumber—Vertical Collector

1 piece 1 × 8—6 feet
2 pieces 1 × 6—8 feet
1 piece 1 × 6—6 feet
2 pieces 1 × 4—8 feet
7 pieces 1 × 2—8 feet
4 pieces ¾ × ¾-inch baluster stock—8 feet
1 piece ½-inch CDX exterior plywood—4 × 8 feet

Cost

Lumber—Horizontal Collector

1 piece 1 × 8—8 feet
2 pieces 1 × 6—8 feet
3 pieces 1 × 4—8 feet
8 pieces 1 × 2—8 feet
4 pieces ¾ × ¾-inch baluster stock—8 feet
1 piece ½-inch CDX exterior plywood—4 × 8 feet

Cost

All the material you will need is standard dimensional lumber. Get #2-grade wood. If you get #3-grade, you will most likely have a lot of problems with knots, warping, and bowing. The temperatures inside the collector will worsen these problems, so get good wood to start with. You may even want to consider going up a grade and getting clear stock. This is an especially good idea for the 1 × 2 material, as it is the most susceptible to damage from knots. The plywood should be exterior grade.

If you cannot find ¾ × ¾-inch baluster stock, you can get the number of pieces you need from an 8-foot length of 1 × 6 material. Have the strips ripped for you, or do it yourself.

Insulation

2 pieces ⅝-inch insulation board—4 × 8 feet

Cost

Solar Air Heater

We recommend you use ⅝-inch solid insulation board. Don't confuse board insulation with weatherboard or sheathing board. Insulation board, which comes in thicknesses up to 2 inches, has a reflective coating on both sides and is made of compressed foam or, in some cases, compacted fiberglass. It should have an R-value of about 5 in the thickness recommended.

Glazing

1 or 2 pieces (number varies) fiberglass-reinforced plastic—4 × 8 feet

The material we have found best for home construction is fiberglass-reinforced plastic. This is normally almost clear with a slight grayish tint. It does not shatter but will crack. It is easy to cut and attach. You can order it by mail or get it at some building supply or greenhouse supply stores. The two major manufacturers of this material are Kalwall Corporation, P.O. Box 237, Manchester, NH 03105 (phone 603-668-8186), and Filon Division, Vistron Corporation, 12333 Van Ness Avenue, Hawthorne, CA 90250 (phone 213-757-5141). These materials are available in different thicknesses; we have found that the thinnest materials work best, as they can be put down with the least amount of wrinkling. The actual thickness of the material we used is 0.025 inch. Illustration 1-3 will show you whether you need one or two layers of glazing.

Hardware—Vertical Collector

SCREWS
2-inch #10 flathead wood screws—14
1¼-inch #8 flathead wood screws—40
⅝-inch #6 flathead wood screws—18

MISCELLANEOUS HARDWARE
³⁄₁₆-inch × 2-inch hanger bolts—6
³⁄₁₆-inch × 2½-inch hanger bolts—18
¼-inch brass flat washers—24
³⁄₁₆-inch brass cap nuts—24
self-closing brass-plated cabinet hinges—3 sets
⅜-inch staples—1 box
1½-inch steel barbed nails—48

Hardware—Horizontal Collector

SCREWS
2-inch #10 flathead wood screws—14
1¼-inch #8 flathead wood screws—42
⅝-inch #6 flathead wood screws—24

MISCELLANEOUS HARDWARE
³⁄₁₆-inch × 2-inch hanger bolts—8
³⁄₁₆-inch × 2½-inch hanger bolts—18
¼-inch brass flat washers—26
³⁄₁₆-inch brass cap nuts—26
self-closing brass-plated cabinet hinges—4 sets
⅜-inch staples—1 box
1½-inch steel barbed nails—48

Miscellaneous

SCREEN
metal screen—48 inches × 17 feet

It is absolutely mandatory that you use a metal screen for your collector screens. Fiberglass window screen will not collect heat. If at all possible get black aluminum screen. Otherwise you will have to paint your screen black.

TAPE
heat-activated duct tape—1 roll

We have found that heat-activated tape is far and away the most permanent and easiest tape to work with. The problem is that this tape is normally produced for commercial use only. You will have to buy a roll from a heating/air-conditioning supply store. The rolls are normally quite large, and you'll end up with extra tape. The quality of this material is well worth the extra tape and cost. We used Johns-Manville Therm-Lock, produced by Johns-Manville Corporation, Ken-Caryl Ranch, Denver, CO 80217 (phone 303-979-1000). It is 3 inches wide, white on one side and reflective on the other, with reinforcing fibers running through it. A standard iron on a high setting will activate the adhesive. Do not use the silver-color duct tape that is adhesive backed. It will deteriorate with the collector's heat.

ADHESIVES
exterior-grade panel adhesive—1 tube
carpenter's white glue—2 ounces

Cost

PAINT
1 quart primer
1 quart flat black
1 quart exterior paint

Cost

WOOD FILLER
1 small can exterior-grade filler

Cost

CAULKING
silicone caulking—1 tube

Cost

WEATHER STRIPPING—VERTICAL COLLECTOR
⅜-inch tubular vinyl gasket—17 feet

Cost

WEATHER STRIPPING—HORIZONTAL COLLECTOR
⅜-inch tubular vinyl gasket—34 feet

Cost

Collector Materials		
Plans Book		
Total Heater Cost		
Installation Materials		
Hired Labor		
Installation Cost		
Installed Heater Cost		
Minus 40% Federal Tax Credit		
Cost of Installed Heater		
Minus Other Energy Credits		
Final Heater Cost		

2 Building the Collector Box

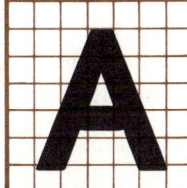

After buying your materials, the first step in building a solar hot air heater is to build the collector box. The box is the actual heat-collecting area of the heater, but it also holds the glazing material in place and is the point at which you attach the heater to your house. As a foundation is to a house, so the box is to the heater—if you build a good, square box, you'll end up with a good, square heater. If your box is off, then the rest of the unit will be off.

The collector box is easy to build. It is really only a sheet of insulated plywood with a frame and a door attached to it. There are a few other small pieces to be attached, but all you really have to do is put a four-sided frame on a piece of plywood.

The entire collector box can be cut out with nothing more complicated than a saber saw or even a coping saw. A power drill will speed the process along greatly. A combination bit that drills a pilot hole, a shank hole, and a countersink all at once will be a big time saver.

The heaters use one full sheet of plywood for the back, 1A or 2A. The rest of the pieces are all cut from standard dimensional lumber. We have designed the heater to use all standard sizes to avoid the need for rip cuts. If you cannot get a certain dimensional size lumber, have some ripped for you. Use chart 2-1 to see the actual size of dimensional lumber. From time to time, the width of dimensional lumber will vary. Be sure all the pieces you get are the same width. They may

differ slightly from those that we used, but as long as all pieces of one size are uniform, you will have no trouble.

We feel it is best to cut all the pieces for the box before beginning assembly, to reduce the number of times you have to paint. However, we do not recommend that you cut the pieces of the glazing frame at the same time as you cut the pieces of the box; they might get mixed up. The glazing frame is explained in the next chapter and on blueprint sheets 3 and 4.

This chapter is divided into two sections; first, instructions for a vertical unit, then, in-

Nominal vs. Actual Lumber Size

Nominal Size	Actual Size
1 × 2	3/4" × 1 1/2"
1 × 4	3/4" × 3 1/2"
1 × 6	3/4" × 5 1/2"
1 × 8	3/4" × 7 1/4"

Chart 2-1—Order your lumber by the nominal size. If you can't get it, have it cut to the actual size.

structions for a horizontal unit. The assembly of the two units is almost identical, but there are a few minor, but extremely important, differences. Read only the section that applies to the type of unit you are building; otherwise, you may get confused. Now let's get started building the collector box for your solar hot air heater.

VERTICAL UNIT

If you are building a vertical hot air heater, read this section of the chapter, blueprint sheet 1, and the vertical cutting diagram on blueprint sheet 6. If you are building a horizontal unit, read the back section of this chapter.

The first step is to cut the pieces, using the vertical cutting diagram on blueprint sheet 6 for the most economical use of materials. At this time, cut only those pieces that are not shaded on the blueprint sheet; the shaded pieces will be cut next chapter. You should cut all the pieces listed on blueprint sheet 1, plus four screen-mounting strips, 3F, and the vent door, 3E, although they will not be assembled at this time. They should be primed and painted now; you will assemble them later.

The back, 1A, does not have to be cut; it is a full 4-foot by 8-foot piece of plywood. The first piece to cut is the top, 1B. This should be cut from a length of 1 × 8 material. Cut the top exactly 48 inches long. Be sure the factory end is square before marking the piece for your cut, and be sure you make your cut square.

Next, cut the two sides, 1C, from 1 × 6 material. The sides should be cut off square, 94½ inches long, unless your wood is of a thickness other than ¾ inch. If so, the sides should be cut off at a length equal to 96 inches minus the combined thickness of the top, 1B, and the bottom, 1D. Be sure the pieces of wood you use for the sides are not warped or bowed at all. The sides must be straight and true.

Cut the bottom, 1D, from a length of 1 × 6 material. The bottom should be 48 inches long. The bottom must have two vent openings cut into it. This is to allow the unit to be cooled during the summer months. Illustration 2-1 shows the placement of these

Illustration 2-1—Vent hole position in the bottom is very important. Note that the vents are off-centered on the width of the board.

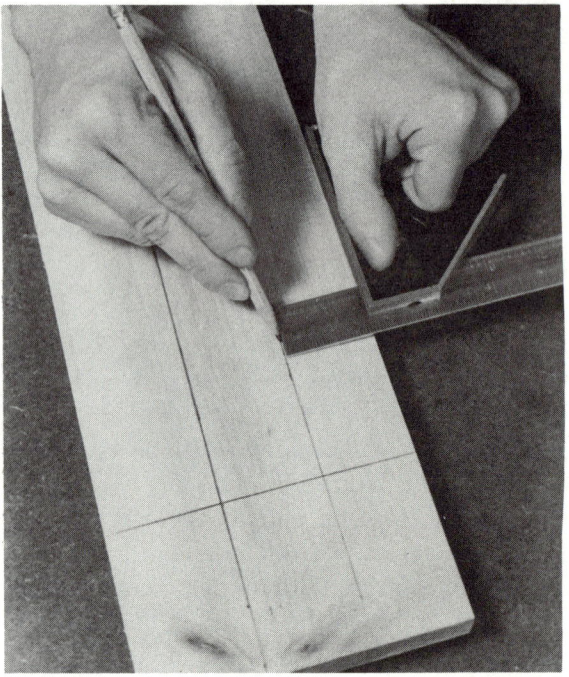

Photo 2-1—Use a combination square as a sliding marking gauge to mark the vent openings, in place of a long straightedge.

vent holes. Note that the vent holes are offset. Measure carefully before marking and cutting them.

If you do not have a long straightedge to mark the openings, use the end of your combination square as a sliding marking gauge, as shown in photo 2-1. To do this, first measure and mark the dimensions on the wood, then adjust the square so a pencil point, when held against the end of the rule, falls on the mark. Slide the square along the wood (keeping it firmly against the edge of the wood), holding the pencil against the end of the rule and marking the dimension along the length of the wood. Be careful of splinters. Then come back and mark the end and center sections, as shown in illustration 2-1.

A saber saw is best for cutting the vent openings. To use the saber saw, drill a hole at each corner of each opening and, with a rip blade in your saw, cut the openings. A rip blade will help cut down the time this cutting takes. This type of cut takes much longer with a crosscut blade. Safety glasses should be worn when making these cuts.

Before cutting any more pieces, carefully inspect your supply of 1 × 2 material. Unless you bought clear wood, you will have to deal with knots and your wood may be warped or bowed. Carefully look at the cutting diagram for the 1 × 2 material, and be sure to save the best areas of these lengths of wood for use in the next chapter; those pieces are shaded. Once you've decided which areas of the wood to save for the glazing frame, cut the cleat, 1E, 46½ inches long.

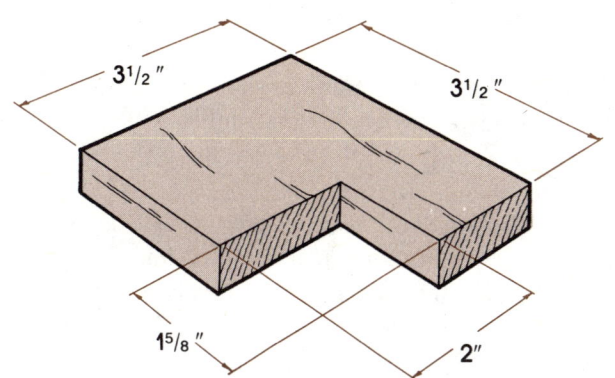

Illustration 2-2—Be sure the grain is in the right direction on the screen-mounting brackets to prevent them from splitting when installed.

From the 1 × 2s, cut the four screen-mounting strips, 3F, 46¼ inches long. Remember, these will not be used now; they will be primed and painted at this stage, but assembled later.

From a length of 1 × 4, cut the four screen-mounting brackets, 1F, 3½ inches long. From the same piece of 1 × 4, cut the four baffles, 1G, 10 inches long. Gather together the four screen-mounting brackets. Look at illustration 2-2 for details of how each piece should be trimmed. All four pieces should be identical. Pay attention to the grain direction on these pieces.

From another length of 1 × 4, cut two identical vent doors, 1J and 3E, each 46 inches long. One door will be installed on the collector box, the other on the glazing frame. Each vent door must have three notches cut in it, as shown in illustration 2-3. To make the notches, drill a ¼-inch hole at the center point and then cut out the notch area with a saw. After cutting the notches, round all edges on the doors with a wood rasp or other shaping device, to give the door a finished look.

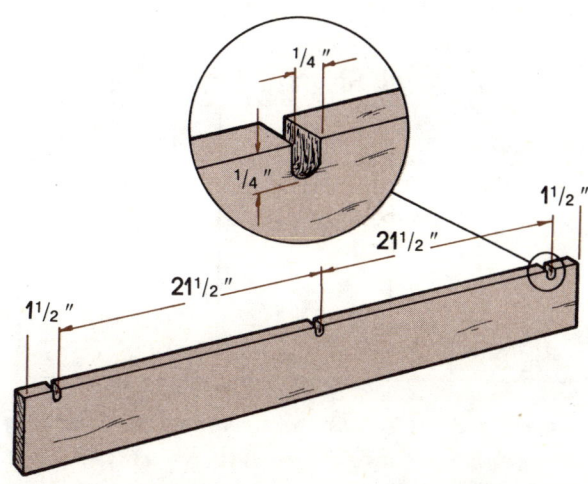

Illustration 2-3—Each vent door must have three notches. The large drawing shows the center points for the starting holes, and the detail shows the finished notches.

Solar Air Heater

Last, from a piece of scrap, as shown on the cutting diagram, trim a 1½-inch by 3-inch piece, 1H, for the thermostat mount.

ASSEMBLY

Before starting assembly, go over each piece and sand and smooth any splinters or rough areas left from cutting. You should break the edge, that is, slightly round the sharp edges, on all pieces to prevent sharp, milled edges from being damaged during assembly and installation. This will give the unit a more professional, finished look.

You will be gluing and screwing all joints during assembly. Chart 2-2 gives recommended pilot and shank hole sizes for the screws used in making your collector. Be sure to use the proper-size pilot hole and to countersink all screws.

Begin by assembling the frame of the collector box. Four pieces will be joined with simple butt joints. Each joint should be glued and screwed with three 2-inch #10 wood screws. Work on a flat floor to keep the edges of the pieces aligned.

The top and bottom pieces overlap the side pieces, as shown in illustration 2-4. Be sure both the bottom and top pieces are in place properly before fastening. The wider portion of the wood of the bottom should be toward the floor, putting the vents away from the floor. The top is wider than the sides and bottom to create an overhang which will keep rain from getting into the exhaust vent. This overhang should also be away from the floor.

We have found it best to hold two pieces together in position, drill a hole for one screw, apply glue to both pieces, and screw them together. Then drill the other two holes and screw them together. Be careful not to overtighten the screws, as they are going into end grain which can be easily stripped of its threads.

With the frame assembled, turn it upside down, so the overlap of the top is toward the floor. Apply panel adhesive to all four edges of the frame, and lay the plywood back in place on top of it. Align the frame and plywood at one corner and nail it with 1½-inch steel barbed nails. Then adjust the rest of the frame to be square with the plywood and nail the other three corners. As you nail the plywood to the rest of the frame, with the nails about 6 to 8 inches apart, be sure the edges of

Pilot and Shank Hole Sizes

Screw	Pilot Hole	Shank Hole
#10	3/32"	3/16"
#8	5/64"	11/64"
#6	1/16"	9/64"
3/16" Hanger Bolt	3/32"	None

Chart 2-2 — Use the recommended drill for each size screw you will be using.

Illustration 2-4 — Be sure the bottom is properly positioned in relation to the top before assembling the box.

the frame stay flush with the plywood's edges. You have to assume the plywood is square and build your collector to fit it.

With the box assembled and attached to the back, fill all openings with exterior-grade wood filler, using a small putty knife. Don't overlook any area. Every inch of all edges of the plywood should be filled. Be sure to seal the end grain of the top and bottom of the frame, as well as any knots or rough surfaces. If you do a good job of sealing all areas now, the finished collector will weather better in the future.

With all the exposed outside surfaces and any big knots or cracks on the inside filled, sand the filler smooth and you will be ready to prime the unit. There is a lot of discussion about what type of primer and paint is best to use on solar devices. We think it best to use exterior-grade primer and paint formulated by the same company, so that the primer will have been blended to be compatible with the paint. For your collector, you'll need the best paint job possible. Don't try to save a few dollars by buying discount paint or by using an old can you found in the garage. Get the best-quality exterior primer and paint you can find. It doesn't matter whether it's latex or enamel, just as long as the primer is designed to be used with the paint.

Prime the box, as well as all the other pieces you have cut to this point. Be sure to put a good coat of primer on all surfaces of the box, inside and outside, and all sides and ends of the small pieces.

When the primer dries, you are ready to begin assembling the inside of the collector box. Begin by cutting the back piece of insulation, 1L. This piece should be cut with a utility knife to measure approximately 46½ inches by 94½ inches. Measure the inside of the frame to be sure of a good fit. The insulation should fit as firmly inside the collector frame as possible. Use a straight length of wood as a cutting guide to get straight edges.

Don't test-fit the whole piece after it is cut. If you cut it properly for a tight fit, you will not be able to remove it from the box after test-fitting. To test the fit of the insulation, try each edge separately in the area where it will go. Try one edge and pull the insulation out; try another edge, and so on. Once you are sure it will fit, apply panel adhesive to the inside back of the box and insert the insulation. Use a length of 2 × 4 to press the insulation firmly in place along the edges without breaking it.

Next, screw the cleat, 1E, to the top, 1B, with four 1¼-inch #8 wood screws. The top edge of the cleat should be flush with the sides of the box, as shown in illustration 2-5. Fasten the ends first, then measure to be sure the middle is at the same height and attach it. This will compensate for any bow in the cleat. Be careful your pilot holes don't go through the top. All you need are small starting holes; the screws will do the rest. For added strength, apply glue to the cleat before screwing it in place.

With the cleat firmly attached, cut and install the top piece of insulation, 1M. This piece of insulation should fit firmly between the bottom of the cleat and the back insulation and be flush from one side of the box to the other. The insulation material must be protected from direct sunlight. To do this, tape any edges that may be exposed to sun-

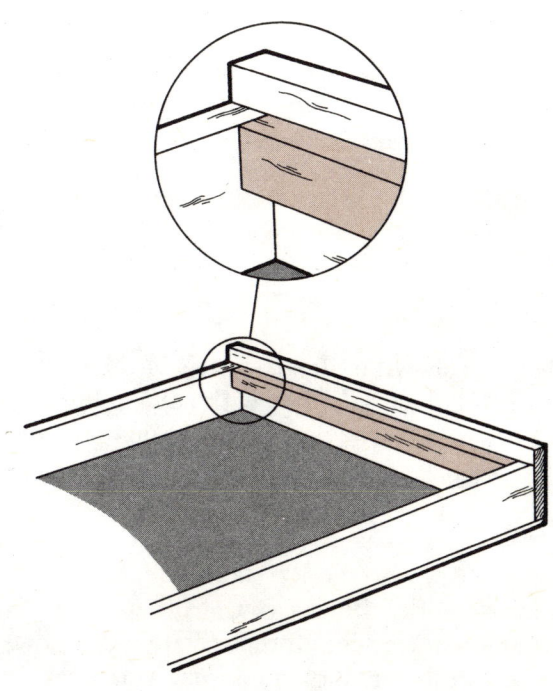

Illustration 2-5—The top edge of the cleat should be flush with the edges of the sides. Be sure to measure the height at the middle of the cleat and correct sagging or bowing before fastening it.

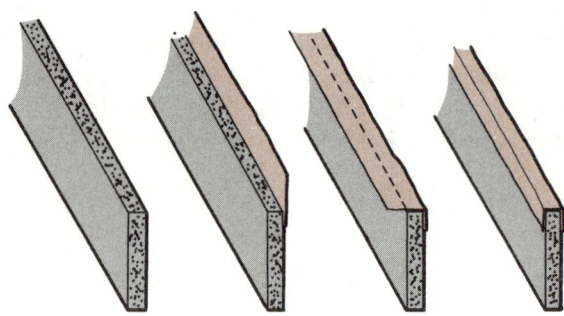

Illustration 2-6—All exposed edges of the insulation must be taped to protect the insulation from sunlight.

light, as shown in illustration 2-6, then apply the insulation to the box with panel adhesive.

As discussed in the materials section, our recommended tape is heat-activated tape. To use this, cut it to length, hold it in place the way you want it, and activate the adhesive by ironing it. This type of tape is easy to work with, as you position it exactly where you want it before activating the adhesive. With contact adhesives, you have to work very carefully, or the tape will stick where you don't want it.

After applying the insulation to the box, attach the four screen-mounting brackets, 1F. Be extremely careful to get the brackets in the proper position, as shown in illustration 2-7. Use two 1¼-inch #8 wood screws for each bracket, and apply glue to the bracket before installing it. Do not drill through the side of the box with the pilot hole, and be sure you have a hole for a screw in each part of the bracket, as shown in photo 2-2.

With the screen-mounting brackets in-

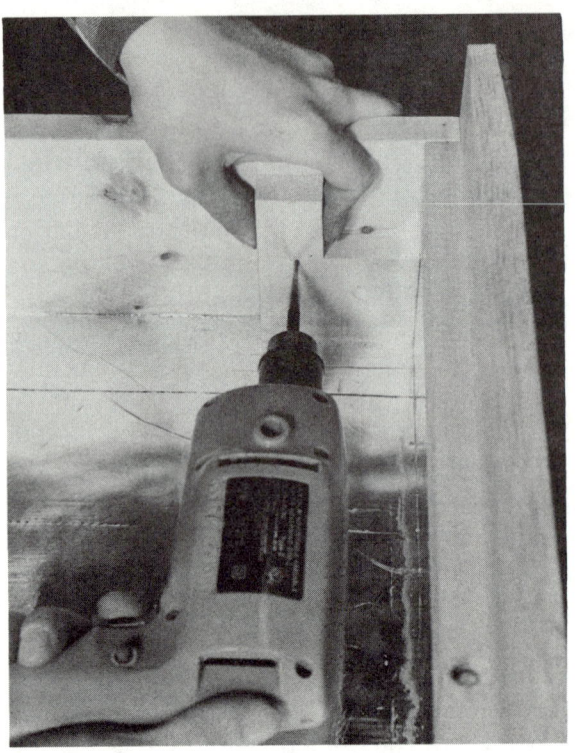

Photo 2-2—The screen-mounting brackets need two holes to fasten them to the box. Note the hole placement to prevent cracking. Don't drill the pilot holes all the way through the side of the box.

stalled, you are ready to measure and cut the two pieces of side insulation, 1N. The side insulation should fit exactly between the two screen-mounting brackets on each side. Measure the space in your box before cutting the insulation. The pieces should be 3¼ inches wide. Be sure to tape the top and side edges before gluing the insulation in place.

Illustration 2-7 shows the placement of the four baffles, 1G. Each baffle should be positioned and the position marked on the back insulation. Then, using a utility knife, cut out the insulation where the baffles will go, as shown in photo 2-3. Glue the baffles in place with panel adhesive, then tape the joints between the insulation and the baffles. Be sure the baffles angle the right way, as shown in illustration 2-7. There is no pressure on the baffles; glue and tape is all that is needed to hold them in place.

The next step is to install the thermostat mount, 1H. This small block of wood is screwed against the side insulation. Position it as shown in illustration 2-7 and on blueprint sheet 1. Use two 2-inch #10 wood screws to fasten it. Drill pilot holes through the mount, being careful not to go all the way through the side. There is very little pressure on the mount; one screw will hold it. However, two screws will insure that, even if the piece cracks, it will stay in place.

At this point the box is complete. Look over the insulation carefully, and if you see any holes or tears, tape them to protect the insulation from sunlight. Tape is most likely to peel up on flat surfaces, so extra care is needed. Clean the area where you will be taping carefully, and press the edges of the tape down well.

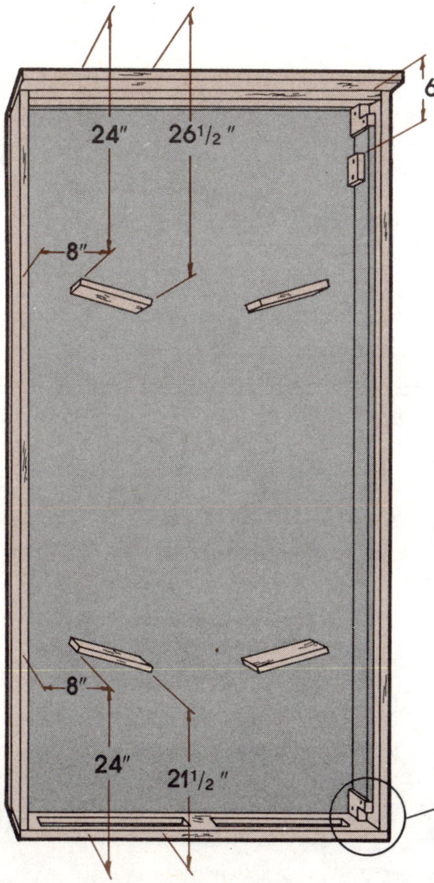

All that remains to be done is to tape the joints between the side, top, and back insulation. Illustration 2-8 shows all areas of the collector box that need to be taped; be sure you don't miss any.

With all the insulation taped, paint the entire inside of the box and the four screen-mounting strips, 3F, flat black. Rub the insulation clean with a cloth before painting. It may take two coats of paint to get a good

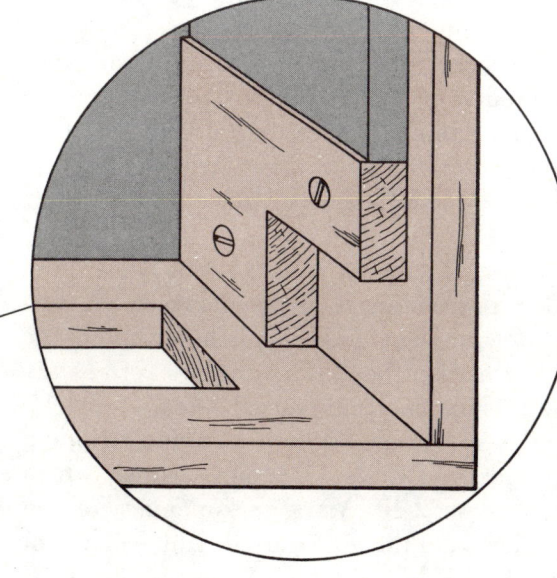

Photo 2-3 — Use a utility knife to remove an area of the back insulation to insert the baffles.

covering over the reflective surface of the insulation.

When the black paint is dry, paint the outside of the collector box whatever color you feel will look best on your house. Be sure to paint only one vent door the same color as the outside of the box. The other vent door

Illustration 2-7 — *The position of the screen-mounting brackets is tricky; be careful to get it correct (see detail). Baffle and thermostat-mount placement are also shown here.*

Solar Air Heater

25

Illustration 2-8—The light-colored areas should be taped when the collector box is assembled.

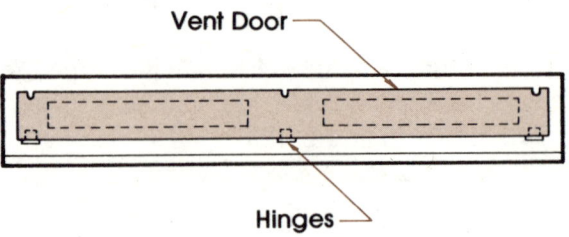

Illustration 2-9—The vent door is attached to the collector box with three hinges. Be sure the door is properly positioned over the vent openings.

will be painted with the glazing frame in the next chapter.

When this paint is dry, attach the painted vent door to the box with three self-closing cabinet hinges. For added weather-proofing, we also weather-strip around the vents and bolt the door in place for the winter months.

First attach the hinges to the back of the door, positioning them as shown in illustration 2-9, then screw the door/hinge unit to the box. Position the door so that it overlaps the vent openings evenly on all sides.

Next, staple 3/8-inch weather stripping around the two vent openings, as shown in illustration 2-10. Mark the openings of the notches with the door closed, and drill a pilot hole for each 3/16-inch by 2-inch hanger bolt. These holes should be positioned toward the closed end of the notch, leaving enough room for the hanger bolt.

The collector box is now completed. Set it aside, and turn to the next chapter to build the glazing frame.

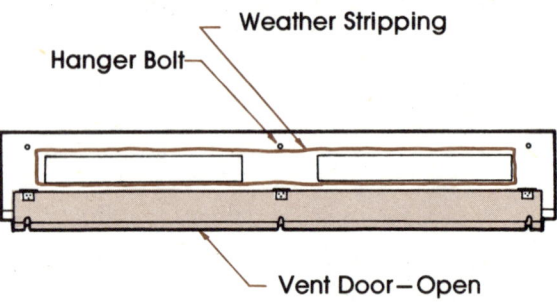

Illustration 2-10—The weather stripping should not interfere with the hanger bolts.

HORIZONTAL UNIT

If you are building a horizontal hot air heater, read this part of the chapter, blueprint sheet 2, and the horizontal cutting diagram on blueprint sheet 6.

Blueprint sheet 2 shows all the pieces in a horizontal heater box. The cutting diagram on blueprint sheet 6 shows all the pieces you

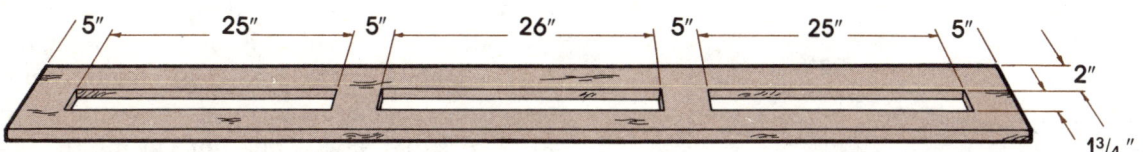

Illustration 2-11—Vent hole position in the bottom is very important. Note that the vents are off-centered on the width of the board.

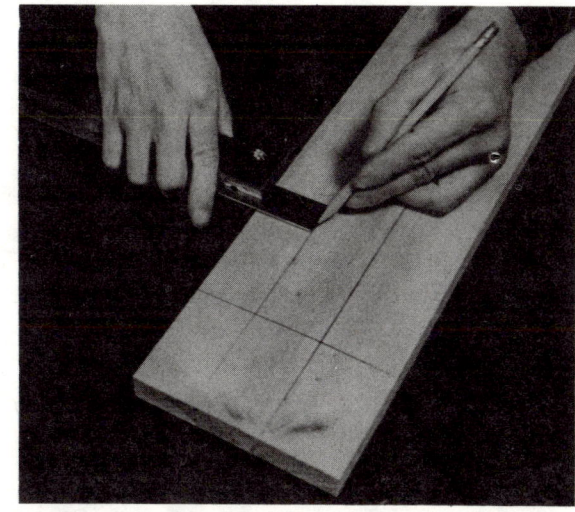

Photo 2-4—Use a combination square as a sliding marking gauge to mark the vent openings, in place of a long straightedge.

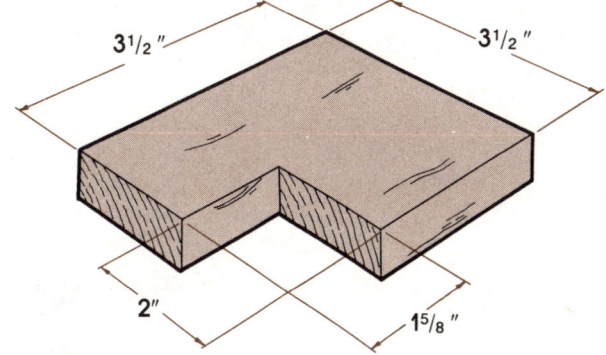

Illustration 2-12—Be sure the grain is in the right direction on the screen-mounting brackets, to prevent the wood from splitting later.

will be cutting at this time, unshaded. Note that five additional pieces from blueprint sheet 4—four screen-mounting strips, 4F, and the vent door, 4E—should be cut at this time. You will cut, prime, and paint those pieces now and assemble them later. The blueprint sheets give the dimensions of all the pieces as well as the nominal size of wood from which they are cut. Be careful to cut each piece from the correct size wood. Blueprint sheet 6 gives the layout of the pieces for the best economy of materials.

The back, 2A, does not have to be cut; it is a full 4-foot by 8-foot piece of plywood. The top, 2B, is a full 8-foot 1 × 8 board.

Cut the two sides, 2C, from 1 × 6 material. The sides should be cut 46½ inches long, unless your wood is of a thickness other than ¾ inch. If so, the sides should be cut off at a length equal to 48 inches minus the combined thickness of the top, 2B, and the bottom, 2D.

For the bottom, 2D, use a full 8-foot 1 × 6 board. The bottom must have three vent openings cut into it. This is to allow the unit to be cooled during the summer months. Illustration 2-11 shows the placement of the vent holes in the bottom. Note that the vent holes are offset. Measure carefully before marking the vent openings.

If you do not have a long straightedge to mark the opening, use the end of your combination square as a sliding marking gauge, as shown in photo 2-4. To do this, first measure and mark the dimensions on the wood, then adjust the square so a pencil point, when held against the end of the ruler, falls on the mark. Slide the square along the wood (keeping it

Solar Air Heater

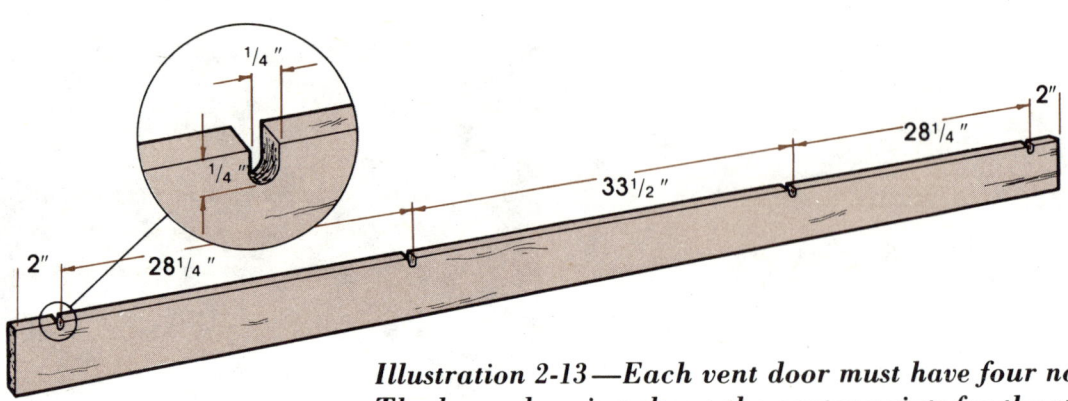

Illustration 2-13—Each vent door must have four notches. The large drawing shows the center points for the starting holes, and the detail shows the measurements of the notches.

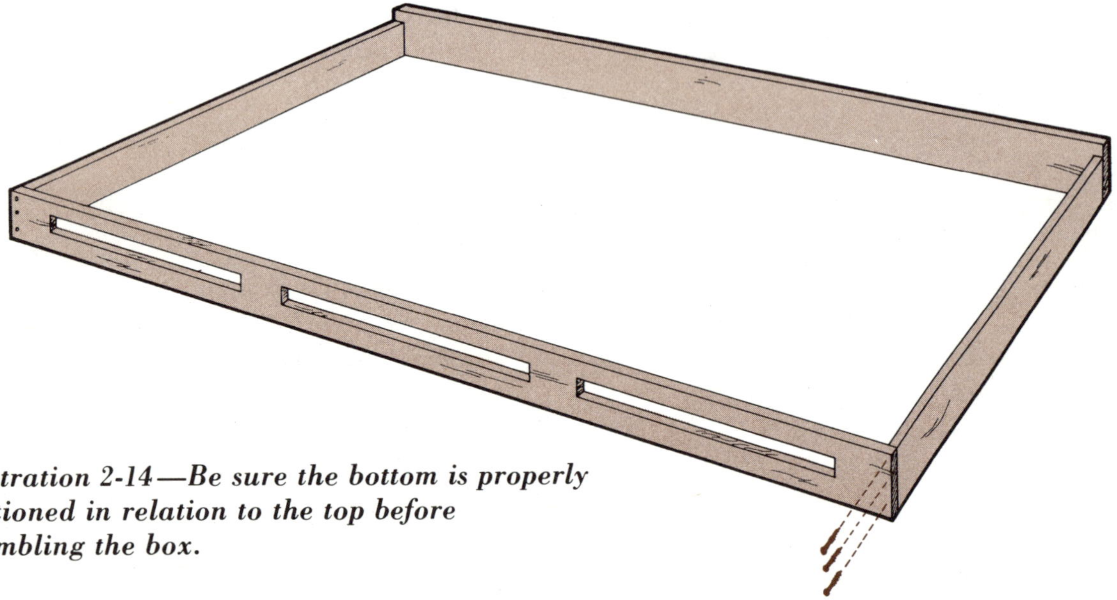

Illustration 2-14—Be sure the bottom is properly positioned in relation to the top before assembling the box.

firmly against the edge of the wood), holding the pencil against the end of the rule and marking the dimension along the length of the wood. Be careful of splinters when doing this. Then come back and mark the end and center sections, as shown in illustration 2-11.

A saber saw is best for cutting the vent openings. To use the saber saw, drill a hole at each corner of each opening and, with a rip blade in your saw, cut the openings. A rip blade shortens the amount of time this cutting takes. This type of cut takes much longer with a crosscut blade. Be sure to wear safety glasses while cutting.

Before cutting the rest of the pieces, carefully inspect your supply of 1 × 2 material. Unless you bought clear wood, you will have to deal with knots, and your wood may be warped or bowed. Look carefully at the cutting diagram for the 1 × 2 material, and be sure to save the best pieces of this wood for the shaded pieces on the cutting diagram that will be used later in building the glazing frame. Once you've looked over the wood and decided which pieces to use now, cut the cleat, 2E, 86¼ inches long.

From four 8-foot lengths of 1 × 2, cut the four screen-mounting strips, 4F, each 46¼ inches long. Remember, these will not be used now; they will be primed and painted at this stage, but assembled later.

From a length of 1 × 4, cut the four screen-mounting brackets, 2F, each 3½ inches long. From the same piece of 1 × 4, cut two baffles, 2G, one 18 inches long and the other 24 inches long. Look at illustration 2-12 for details of how each screen-mounting

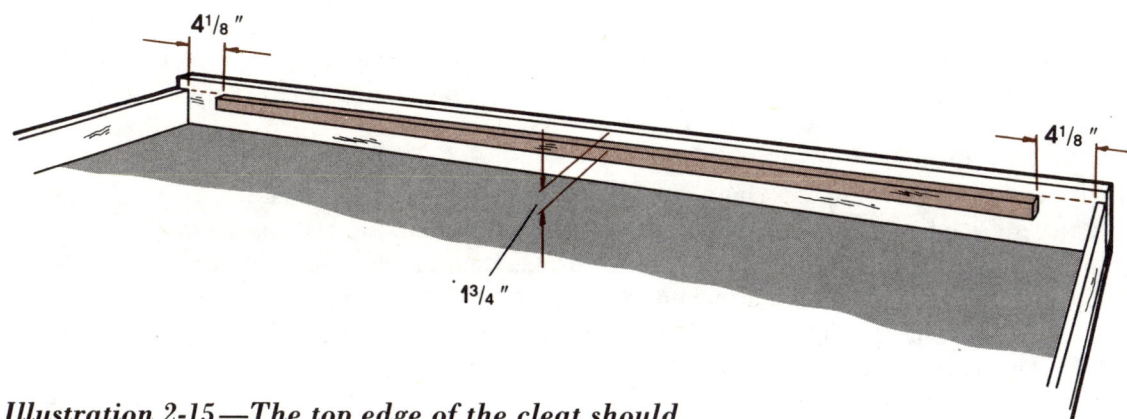

Illustration 2-15—The top edge of the cleat should be flush with the edges of the sides. Be sure to measure the height at the middle of the cleat and correct sagging or bowing before fastening it.

Photo 2-5—Use a combination square to align the ends and middle of the cleat. Be sure the middle is properly positioned, as shown in the detail on right.

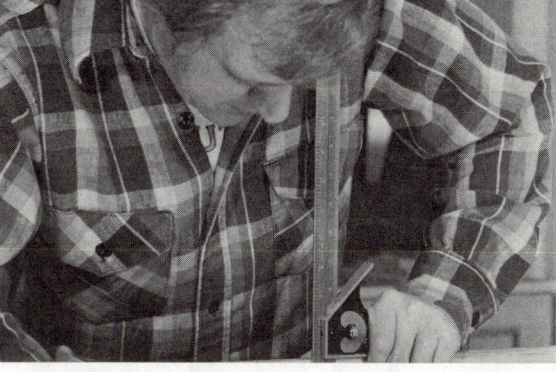

shown in illustration 2-13. To cut the notches, first drill a ¼-inch hole at the center point, and then cut out the notch area with a saw. After cutting the notches, round all edges on the doors with a wood rasp or other shaping device to give the door a finished look.

Last, from a piece of scrap, as shown on the cutting diagram, trim a 1½-inch by 3-inch piece, 2H, for the thermostat mount.

ASSEMBLY

Before starting assembly, go over each piece to sand and smooth any splintered or rough areas left from cutting. You should break the edge, that is, slightly round the sharp edges, on all pieces to prevent the sharp, milled edges from getting damaged during assembly and installation. This will give the finished unit a more professional look.

You will be gluing and screwing all joints during assembly. Chart 2-2 (page 22) gives recommended pilot and shank hole sizes for the screw sizes used in making your collector. Be sure to use the proper-size pilot hole and countersink all screws.

Begin by assembling the frame of the collector box. Four pieces will be joined with simple butt joints. Each joint should be glued and screwed together with three 2-inch #10 wood screws. Work on a flat surface to keep the edges of the pieces aligned.

The top and bottom pieces overlap the side pieces, as shown in illustration 2-14. Be sure both the bottom and top pieces are in the correct position before attaching them to the

bracket should be trimmed and how the grain should run. All four pieces should be identical.

From two other lengths of 1 × 4, cut two identical vent doors, 2J and 4E, each 94 inches long. One door will be installed on the box, the other on the glazing frame. Each vent door must have four notches cut in it, as

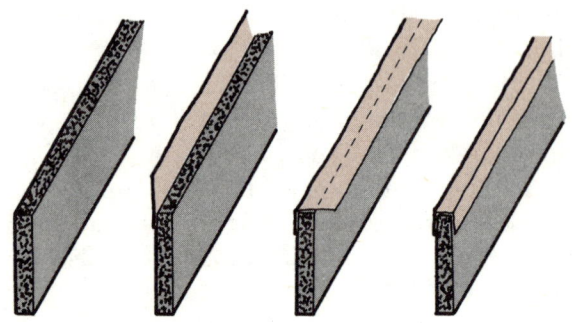

Illustration 2-16—All exposed edges of the insulation must be taped to protect the insulation from sunlight.

sides. The wider portion of the wood of the bottom piece should be toward the floor, putting the vent openings away from the floor. The top is wider than the sides and bottom to create an overhang which will keep rain from getting into the exhaust vent. This overhang should also be away from the floor.

We have found it best to hold two pieces together in position, drill a hole for one screw, apply glue to both pieces, and screw them together. Then drill the other two holes and put in the screws. Be careful not to overtighten the screws, as they are going into end grain, which can be easily stripped of its threads.

With the frame assembled, turn it upside down, so the overlap of the top is toward the floor. Apply panel adhesive to all four edges of the frame, and lay the plywood back in place on top of it. Align the frame and plywood at one corner and nail it with 1½-inch steel barbed nails. Then adjust the rest of the frame to be square with the plywood and nail the other three corners. As you nail the plywood to the frame, with nails every 6 to 8 inches, be sure the edges of the frame stay flush with the plywood's edges. You have to assume the plywood is square and build your collector to fit it.

With the box assembled and attached to the back, fill all openings with exterior-grade wood filler, using a small putty knife. Don't overlook any area. Every inch of all edges of the plywood should be filled. Be sure to seal the end grain of the top and bottom of the frame, as well as any knots or rough surfaces. If you do a good job of sealing all areas now, the finished collector will weather better in the future.

With all the exposed outside surfaces and any big knots or cracks on the inside filled, sand the filler smooth, and you are ready to prime the unit. There is a lot of discussion about what type of primer and paint is best to use on solar devices. We think it best to use an exterior-grade primer and paint formulated by the same company, so that the primer will have been blended to be compatible with the paint. For your collector, you'll need the best paint job possible. Don't try to save a few dollars by buying discount paint or by using old paint you have on hand. Get the best quality exterior primer and paint you can find. It doesn't matter whether it's latex or enamel, just as long as the primer is designed to be used with the paint.

Prime the box as well as all the other pieces you have cut to this point. Be sure to put a good coat of primer on all surfaces of the box, inside and outside, and all sides and ends of the small pieces. Don't forget the four screen-mounting strips and the extra vent door.

When the primer is dry, you are ready to begin assembling the inside of the collector box. Begin by cutting the back piece of insulation, 2L. This piece should be cut with a utility knife to measure approximately 46½ inches by 94½ inches. Measure the inside of the frame to be sure of a good fit. The insulation should fit inside the collector frame as firmly as possible. Use a straight length of wood as a cutting guide to get straight edges.

Don't test-fit the whole piece after it is cut. If you cut it properly for a tight fit, you will not be able to remove it from the box after test-fitting. To test the fit of the insulation, try each edge separately in the area where it will go. Try one edge, and pull the insulation out; try another edge and so on. Once you are sure it will fit, apply panel adhesive to the inside back of the box, and insert the insulation. Use a length of 2 × 4 to press the insulation firmly in place along the edges without breaking it.

Next, screw the cleat, 2E, to the top with six 1¼-inch #8 wood screws. The top edge of the cleat should be flush with the sides of the box and spaced as shown in illustration 2-15. To align the ends of the cleat with the edges of the frame, use a square, as shown in photo 2-5. Align and fasten the ends first, then measure to be sure the middle aligns, and attach it. This will compensate for any bow or sag in the cleat. Be careful your pilot holes don't go

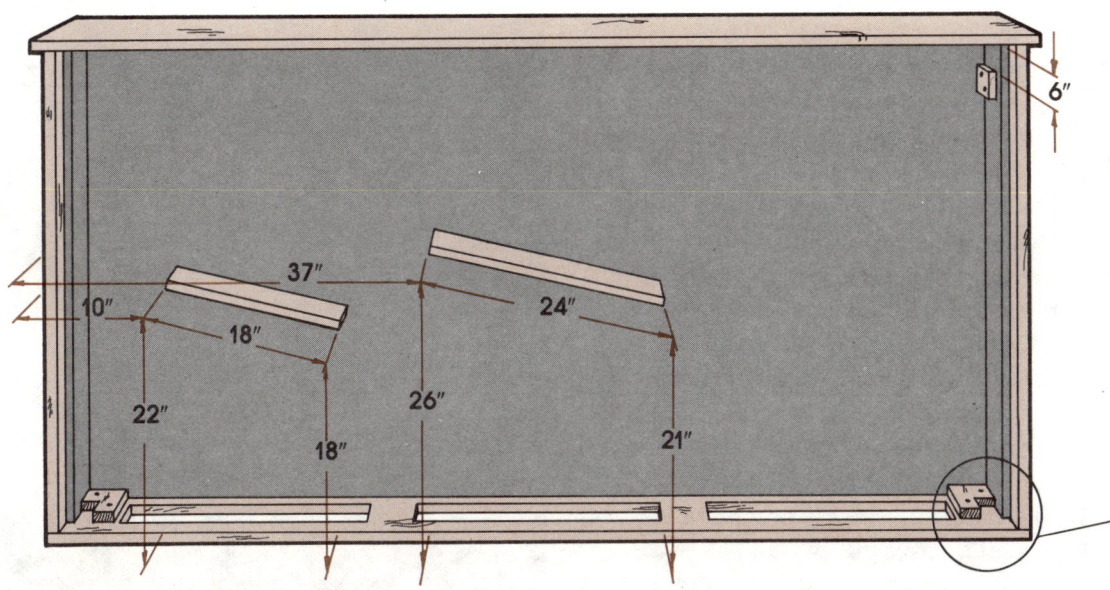

Illustration 2-17—The position of the screen-mounting brackets is tricky (see detail). Baffle and thermostat-mount placement are also shown here. Be sure to get the right-size baffle in the correct position.

through the top. All you need are small starting holes, and the screws will do the rest. For added strength, apply glue to the cleat before screwing it in place.

With the cleat firmly attached, measure, cut, and install the side pieces of insulation, 2N. These pieces of insulation should fit firmly between the top and bottom of the box. Before installing the insulation, tape the top edge and ends of both pieces to protect the insulation from direct sunlight, as shown in illustration 2-16, then apply the insulation to the box with panel adhesive.

As discussed in the materials section, our recommended tape is heat-activated tape. To use this, cut it to length, position it the way you want it, and iron it. The heat of the iron activates the adhesive. This type of tape is easy to work with, as you position it exactly where you want it before activating the adhesive. With contact adhesives, you have to work very carefully, or the tape will stick where you don't want it.

After applying the insulation to the box, attach the four screen-mounting brackets, 2F. Be extremely careful to get the brackets in the proper position, as shown in illustration 2-17. Use two 1¼-inch #8 wood screws for each bracket, and apply glue to the brackets before installing them. Photo 2-6 shows the placement of the second screw for installing the brackets. Be sure that you do not drill through the side of the box with the pilot hole, and that you have a hole for a screw in each part of the bracket.

With the screen-mounting brackets installed, you are ready to measure and cut the top insulation, 2M. The top insulation should be cut to fit flush between the screen-mounting brackets and between the bottom of the cleat and the back insulation. Be sure to tape the top edge and ends before gluing the insulation in place with panel adhesive.

Illustration 2-17 shows the proper placement of the two baffles, 2G. Each baffle

SOLAR AIR HEATER

31

should be positioned and its position marked on the back insulation. Then, using a utility knife, cut out the insulation where the baffles will go, as shown in photo 2-7. Glue the baffles in place with panel adhesive, then tape the joints between the insulation and the baffles. Be sure the baffles angle the right way, as shown in illustration 2-17. There is no pressure on the baffles; glue and tape is all that is needed to hold them in place.

The next step is to install the thermostat mount, 2H. This small block of wood is screwed against the side insulation. Position it as shown in illustration 2-17 and on blueprint sheet 2. Use two 2-inch #10 wood screws to fasten it. Drill pilot holes through the mount, being careful not to go all the way through the side. There is very little pressure on the mount; one screw will hold it. However, two screws will insure that, even if the piece cracks, it will stay in place.

At this point, the box is complete. Look over the insulation carefully, and if you see any holes or tears, tape them to protect the insulation from sunlight. Tape is most likely to peel up on the flat surfaces, so extra care is needed. Carefully clean the area where you will be taping, and press the edges of the tape down well.

All that remains to be done is to tape the joints between the side, top, and back insulation. Illustration 2-18 shows all areas of the collector box that need to be taped; be sure you don't miss any.

With all the insulation taped, paint the entire inside of the collector box and the four screen-mounting strips, 4F, flat black. Rub the insulation clean with a cloth before painting. It may take two coats of paint to get a good covering over the reflective surface of the insulation.

When the black paint is dry, paint the outside of the collector box whatever color you feel will look best on your house. Be sure

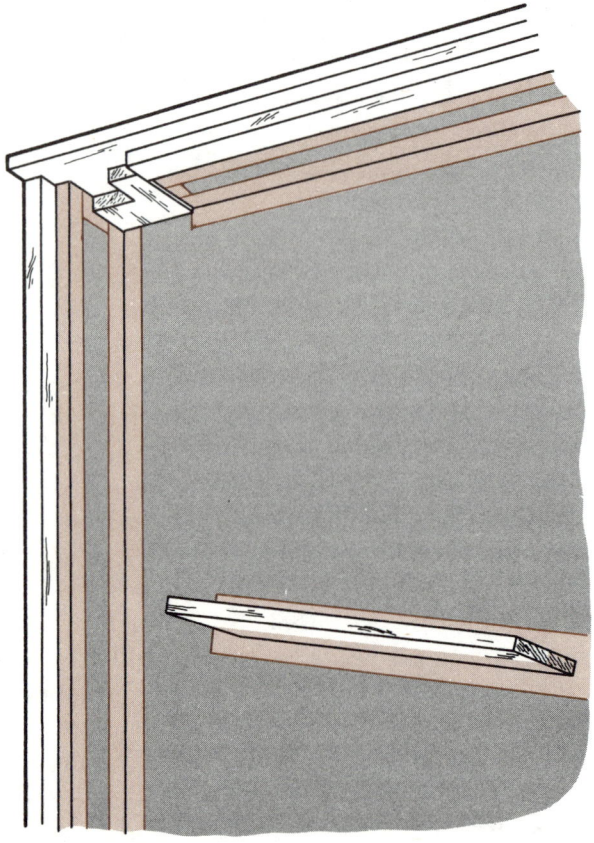

Illustration 2-18—The light-colored areas should be taped when the collector box is assembled.

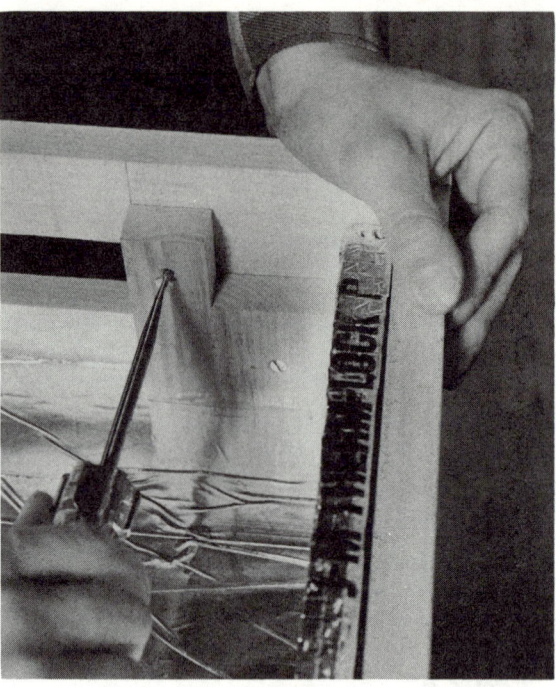

Photo 2-6—The screen-mounting brackets need two screws to prevent splitting. Position the screws as shown.

to paint only one of the vent doors at this time. The other vent door will be painted with the glazing frame, in the next chapter.

When this paint is dry, attach the painted vent door to the bottom of the box with four self-closing cabinet hinges. For added weatherproofing, we also weatherstrip around the vents and bolt the door in place for the winter months.

First attach the hinges to the back of the door, positioning them as shown in illustration 2-19, then screw the door/hinge unit to the box. Be sure the door is properly positioned over the vent openings, as shown in illustration 2-19.

Next, staple a strip of ⅜-inch weather stripping around each vent opening, as shown in illustration 2-20. Mark the openings of the four notches with the door closed, and drill a pilot hole for each $^3/_{16}$-inch by 2-inch hanger bolt. These holes should be positioned toward the closed end of the notch, leaving enough room for the hanger bolt.

The collector box is now completed. Set it aside, and turn to the next chapter to build the glazing frame.

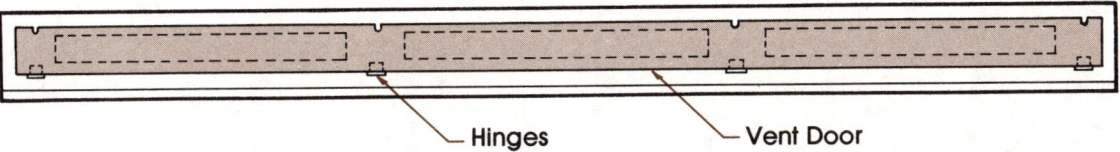

Illustration 2-19—The vent door is attached with four hinges, as shown. Be sure the door is properly positioned over the vent openings.

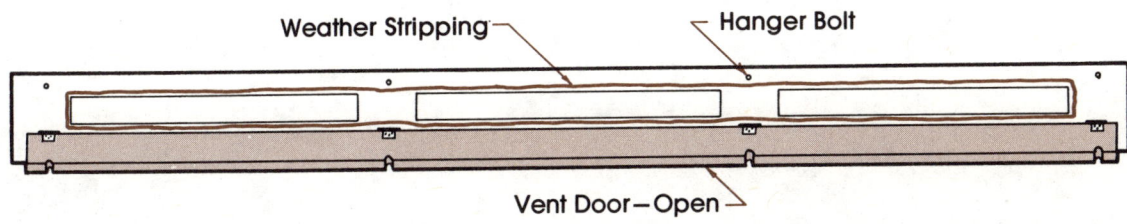

Illustration 2-20—The weather stripping should not interfere with the hanger bolts.

Photo 2-7—Use a utility knife to remove the back insulation for baffle placement.

3 Building the Glazing Frame

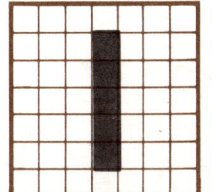

In the last chapter you built the collector box; now you will build and paint the second part of your heater, a glazing frame to cover the collector box. In the next chapter you will install the completed heater.

In this chapter, first you will find instructions for the vertical glazing frame, then for the horizontal glazing frame. Both frames are designed and built essentially the same, but there are differences. Follow only the instructions for the type of unit you are building; if you read the instructions for the other unit, you may get confused.

If you are building a vertical heater, use the first part of this chapter, blueprint sheet 3, and the vertical cutting diagram on blueprint sheet 6.

If you are building a horizontal heater, use the second part of this chapter, blueprint sheet 4, and the horizontal cutting diagram on blueprint sheet 6. On the cutting diagram, the pieces you will be cutting for this chapter are shaded. Several of the pieces you will need were cut in the last chapter; they should already be primed and awaiting assembly.

As pointed out in the materials section, we only use standard dimensional lumber. The entire glazing frame is built from 1 × 2s and a type of trim known as baluster stock. The actual measurement of a 1 × 2 should be ¾ inch thick by about 1½ inches wide. If the wood you get is slightly wider than 1½ inches, it will work fine; just be sure it is all of a uniform width. If you cannot get 1 × 2 material, you can have pieces ripped to 1½ inches from wider stock. Unfortunately, the quality of 1 × 2 material generally leaves a lot to be desired. The wood is so narrow that, if it has any knots, it quickly warps and bows. For the pieces you will be using, some warp and bow can be tolerated, as the glazing frame will be bolted to the collector box to help hold it in shape. However, you will need two pieces that are relatively knot-free. As noted last chapter, be very careful when cutting your 1 × 2s, and be sure to select carefully the area of wood from which each piece is cut. These pieces are very unforgiving, but luckily they don't cost too much, so some error and wasted wood can be tolerated.

Baluster stock is a type of trim carried by most lumberyards. It should measure an exact ¾ inch by ¾ inch. If you are using two layers of glazing, these pieces are used for spacers between them. If you cannot find this material, you can rip ¾-inch by ¾-inch spacers from an 8-foot length of 1 × 6 material. If you don't want to do the ripping, have your lumberyard do it for you. Normally there is a charge for ripping, but it varies greatly from yard to yard. Some will do it free

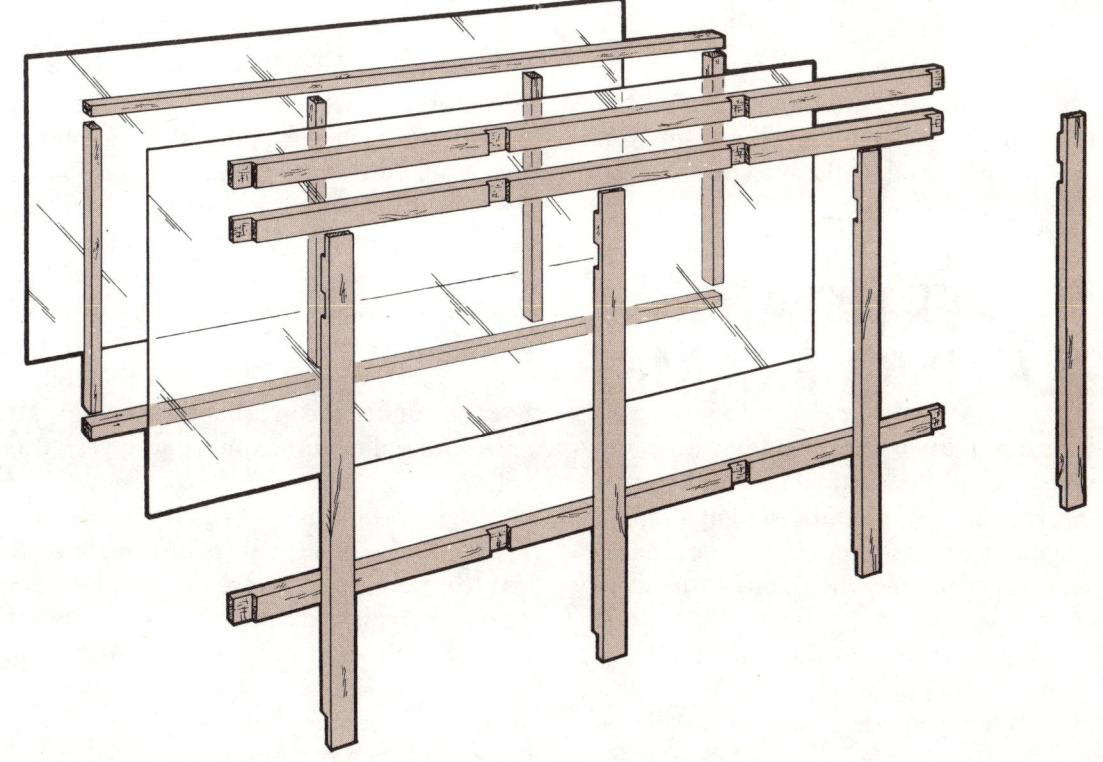

Solar Air Heater

with a large order. This service is often not available at home centers, so ask around before buying your materials, and try to get everything at one place.

It is most important to fill, prime, and paint all wood in the glazing frame well. It is equally important to attach the glazing material to the frame properly and caulk well. Even a small amount of air infiltrating the glazing frame will hurt the efficiency of your solar heater. What you are about to build is not difficult, but there are some tricky steps. Pay careful attention to the details pointed out in this chapter, double-check your measurements before cutting, and you should have no problems. Work slowly, and keep this manual by your side.

Those of you who are building a vertical collector, continue reading; those of you building a horizontal collector, turn to the back of the chapter.

VERTICAL GLAZING FRAME

Use this part of the chapter with blueprint sheet 3 and the vertical cutting diagram on blueprint sheet 6. The first step in building your glazing frame is to cut the pieces. Then they will be assembled and painted, the glazing material will be attached, and the glazing frame will be attached to the collector box.

Look at blueprint sheet 3 for a few minutes. What you are about to build is a frame to cover the opening of the collector

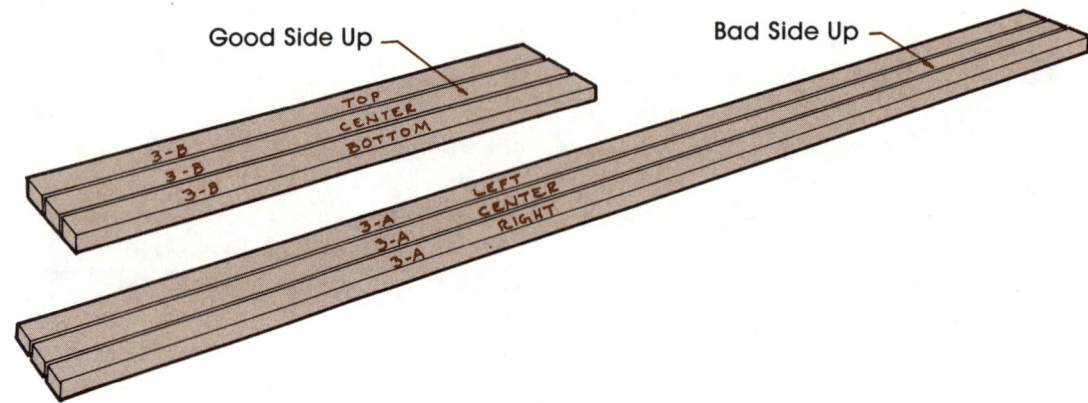

Illustration 3-1—Lay out your pieces as shown. Be especially careful to label the correct side.

box you have already built. If you built your collector box well, the measurements presented here will work perfectly. If you made any modifications, or had problems, check the overall outside edge measurements of the collector box to be sure the glazing frame will fit. The outside-to-outside measurements of the glazing frame should be 48 inches wide and 95¼ inches long. The frame is designed to fit exactly over the frame of the collector box, but under the overhang of the top. Measure your collector box to be sure your frame will fit.

Begin by cutting the six frame members. There are three vertical frame members, 3A, and three horizontal frame members, 3B. These are all cut from 1 × 2 stock. Go through your supply of remaining 1 × 2s, and select good sections of the short and long lengths. From the short lengths, cut three horizontal frame members, 3B, exactly 48 inches long. Be sure both ends of these pieces are square. Then, from your remaining 8-foot lengths of 1 × 2 material, cut three vertical frame members, 3A, 95¼ inches long. You will only have a small amount of scrap on these pieces, so trim the ends to be square carefully, without making the pieces too short.

With all six pieces cut, you have to label them. This may sound like a trivial step, but in reality it is extremely important. If you label your pieces properly, the cutting of the half-lap joints and fitting together of the frame will be very easy, almost foolproof. Pay careful attention to the following instructions.

Lay out all six pieces side by side, the three long pieces together and the three short pieces together. Select the one best long and the one best short piece; these will be your center pieces. The two center members are the most critical, as they do not get bolted to

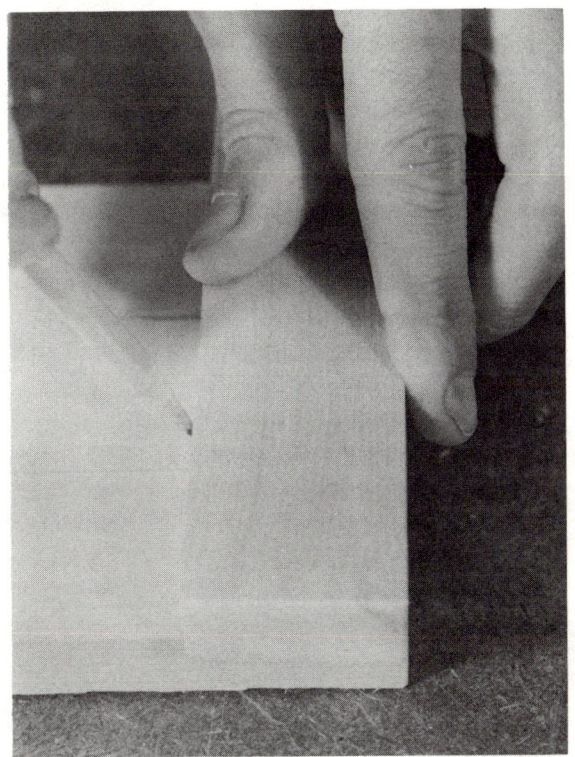

Photo 3-1—Keep the edges all straight and aligned with the marking piece before marking the joint area.

the two center members laid out, put the two other members next to each center member; again, position the short pieces with the good side up, the long pieces with the good side down.

At this point, you should have two piles of wood, one of three short pieces with their good sides up and the best piece, selected as the center piece, flanked by the top and bottom pieces. Label each piece as shown in illustration 3-1. You should have a second pile of three longer pieces, good sides down, with the best piece, selected as the center piece, flanked by the left and right side pieces; label these as shown in illustration 3-1.

Now you are ready to mark each piece for the joint areas. The glazing frame is assembled by cutting a half-lap joint at each end of each of these six pieces and then cutting a center half-lap in each piece, as you will see in a few minutes.

First, we will mark the half-laps on the long, vertical members. Keep the three pieces together, and lay the short, horizontal frame member labeled *top* across one end of all three vertical members, as shown in photo 3-1. Keep the edge of the horizontal member exactly flush with the three ends of the vertical members, and mark the width of the horizontal member across all three vertical members, as shown. The joint areas just marked should be labeled *top*.

Return the top horizontal member to its original pile of pieces, select the horizontal piece labeled *bottom*, and repeat the marking procedure on the other end of the vertical members; these joints should be labeled *bottom*.

To mark the center joint area on the three vertical members, keep them together, and measure 1¾ inches down from the bottom of the top half-lap joint on each piece. Align the top edge of the center horizontal

the collector frame and should be the straightest and strongest two pieces you have. Check to be sure they are not only knot-free, but also straight.

With the two center members selected, lay them out on a flat surface. Lay the short member with its best face up and the long member with its best face down. This is extremely important as you will be cutting all your joints on the sides that are labeled. With

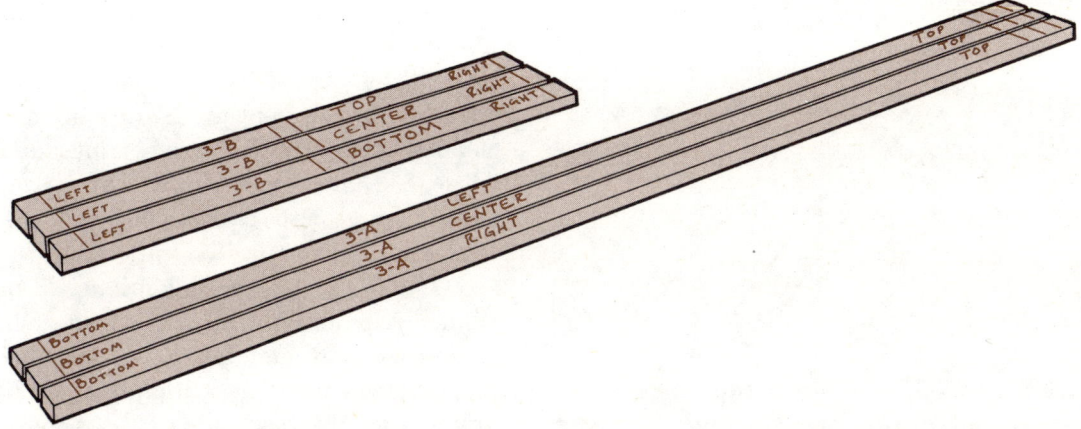

Illustration 3-2—After marking, all pieces of the glazing frame should look like this.

Solar Air Heater

piece with these marks, and mark its width across all three vertical members. (This will require two pencil lines.) When finished, the back of the vertical members should be marked as shown in illustration 3-2. Remember, all markings on the vertical members should be on the bad side of the wood.

Return the three horizontal members to their original grouping, as shown in illustration 3-1. Take the long, vertical member labeled *left*, align it with the left ends of the three horizontal pieces, and mark its width, as you did on the vertical members. Label these joint areas *left*. Repeat with the right vertical piece, on the other ends of the horizontal members, and label the joint areas *right*.

All that remains is to mark the center joint area. Illustration 3-3 shows the measurement for this marking. Measure from one end of each horizontal piece to get the position for the vertical center piece. Then lay the actual piece on these alignment marks, and mark its thickness, as you did for the center joint on the three vertical members. By measuring from one end only, you will compensate for any width difference between our wood and yours. If you have wood of a width other than 1½ inches, your center frame member will be slightly off-center, but you will never notice it. When you are done marking all six frame members, check to be sure your markings match those on illustration 3-2.

HALF-LAPS

The next step is to cut the half-lap joints you just marked. All parts of the glazing frame are assembled with half-laps. Using this method, the outside surface of the frame will be perfectly smooth when assembled.

A half-lap joint is made by removing half the thickness of both pieces of wood at the area to be joined. The two pieces are then attached at the joint area with glue and two small screws. This is a strong joint, due to the increased amount of gluing surface and the interlocking of the two pieces of wood with screws, as shown in illustration 3-4.

Once you get the technique down and your equipment set up, cut all the joints at one session. Remember to cut all pieces on the labeled side. This will produce a finished frame with the best face of the wood facing out.

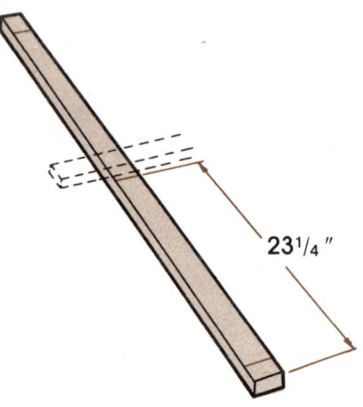

Illustration 3-3—To mark the center joint, measure from one end only, in case your wood is not exactly 1½ inches wide.

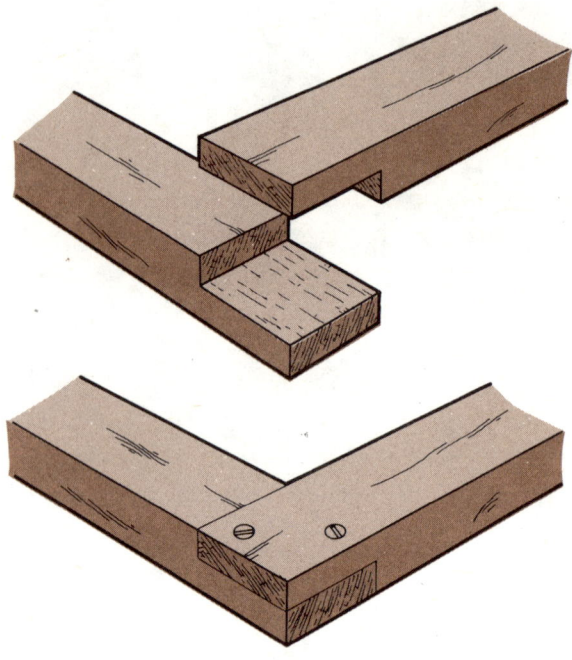

Illustration 3-4—Half-lap joints are strong and easy to cut. Be sure to glue and screw each joint for added strength.

There are several ways to cut half-lap joints. If you have used a radial arm or table saw, most likely you have made half-lap joints before. All you have to do is set the saw for exactly half the thickness of your wood, and cut within the lines you've marked on the pieces. The only thing to watch out for, once you are set up, is that the saw does not vibrate loose and move to a deeper setting.

If you don't have a radial arm saw, don't worry; you can make joints that are just as good with a circular saw, or even a handsaw, but it will take a bit longer. If you have a

Photo 3-2—To cut a half-lap with a circular saw, set the depth, and then cut away the wood within the joint area with successive passes with the saw.

circular saw, you'll cut the joints the same as on a stationary power saw. First, set the depth of your circular saw to half the thickness of your wood. Check your setting by cutting a couple of joints on scrap pieces of wood. Be sure the scrap is the same thickness as the frame members you'll be joining together. Once you have the depth properly set, you can either make many cuts with the saw to remove all the wood, or leave small strips of wood between saw cuts, as shown in photo 3-2, and chisel them off later. Work slowly, and wear safety glasses. It is far better to make the joint too narrow at first, then cut out some additional wood, than to cut too much wood the first time and end up with a weak joint.

When checking a joint, test it with the exact piece of wood that will be joined to it. Making good, tight, smooth joints is worth spending a little extra time. Loose joints are susceptible to moisture seeping into the joint, eventually rotting the wood.

If you don't have any power tools and have to use a handsaw to cut the joints, work slowly, removing only a little wood at a time. Keep the area where wood is removed level and uniform as you go. Don't try to remove all of the joint in one cut-and-chisel effort. Work at it slowly, checking often, and you'll end up with joints as good as those made with power tools. Illustration 3-5 shows the areas to be concerned with when cutting half-lap joints. Check each area as you go to be sure you end up with a tight joint.

ASSEMBLING THE FRAME

With all 18 half-laps cut, the next step is to put the frame together by gluing and screwing each joint. The entire frame must be assembled at one time to get it square, yet you have to work quickly to prevent the glue from drying.

The best way to save time during assembly is to predrill two holes at each joint. These holes should only go in the unlabeled side of the horizontal frame members, the short pieces. Each joint will be fastened with two 5/8-inch #6 wood screws. The size screw requires a 9/64-inch hole. Drill the holes in a diagonal position at each joint, as shown in illustration 3-4. Slightly countersink each hole, so the head of the screw will pull flush with the wood when tight. You do not need pilot holes in the vertical frame members, as this size screw will quickly work its way into them without the aid of a pilot hole.

To assemble the frame, lay the two side vertical frame members, with their labels facing up, on a flat surface. Then test-position the top and bottom horizontal frame members between the two vertical members.

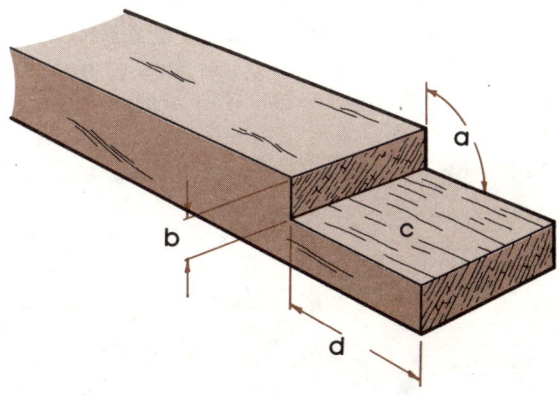

Illustration 3-5—When cutting a half-lap, check a, the angle of the joint; b, the depth of the joint; c, the levelness of the cut area; and d, the size of the cutout area.

Once you have the four pieces properly positioned and are sure they fit together well, apply glue to both faces of the four corner joints and screw the frame members together. The best way to make your frame square is to use the back of the already assembled collector box as your work surface. The glazing frame and the collector box should align exactly. If you work on the back of the collector, once all the pieces align with the plywood, tighten both screws at each joint.

If you don't work on the back of the collector box, put only one screw in each joint, and make them just snug, not tight. Then measure the distance diagonally from corner to corner to see if the frame is square. If the measurements are the same, you've got a square frame; if not, you've got to adjust it. Use your combination square to check each corner until it is square, and then remeasure it for overall squareness. The frame must be in square before you tighten the screws. With the screws tight, remeasure for square, and put the second screw in each joint.

Once you have the four outside pieces of the frame correctly assembled, glue and screw the center frame members in place. At this point, your frame should be firmly assembled. Let it set for a few hours for the glue to dry.

Next, if you are using two layers of glazing, cut the frame spacers. These are simply five ¾-inch by ¾-inch pieces of wood used to keep the two layers of glazing apart. As noted earlier, baluster stock is the ideal size, but if you can't buy that, you can either rip the strips yourself from a 1 × 6 or have it done for you. You need three vertical spacers, each measuring 89 inches, and two horizontal spacers, each measuring 46¼ inches.

With the frame assembled and the spacers cut, prime and paint all pieces. If you did not prime the vent door, 3E, in the last chapter, do so now. As noted earlier, use a primer that is formulated to go with the exterior paint you will use.

The glazing frame and vent door can be painted any color you want, although theoretically, white is the best color. White will reflect some sunlight into the heater, while a darker color would absorb the light, reducing the effectiveness of the heater very slightly.

Before painting, be sure all knots, cracks, and spaces at the joints are caulked with an outdoor wood filler. You want the finished frame to be as smooth as you can get it. Don't forget to fill the screw holes in the back of the frame.

While the primer and paint are drying, attach the collector screen to the screen-mounting strips. The screen serves to increase the amount of absorbent surface inside the collector box without greatly inhibiting airflow. If you can find it, GET BLACK ALUMINUM WINDOW SCREEN. DO NOT GET FIBERGLASS SCREEN; it will not collect heat. If you cannot find black screen, get regular aluminum insect screen, and paint it black. Painting screen is a real time-consuming chore, so check around for black aluminum screen; see if someone can order it for you.

The collector uses two pieces of screen, each approximately 45 inches by 96 inches. The extra screen should be stapled over the vent openings on the inside of the box. Staple the screen well, so insects can't crawl under the screen and into the box. The two 96-inch-long layers are attached together between two screen-mounting strips, 3F, at each end. Lay out the two pieces of screen, one on top of the other, and staple them to one of the screen-mounting strips with ⅜-inch staples. Then, screw another screen-mounting strip to the first with five 1¼-inch #8 wood screws, sandwiching the two layers of screen between two screen-mounting strips, as shown in illustration 3-6.

With one end of the collector assembled, put it into a pair of screen-mounting brackets in the collector box. At the other end of the collector box, put another screen-mounting strip in the brackets. Stretch the screen as tight as you can from one end to the other, and staple the screen to the mounting strip. Then screw the last mounting strip to the other, making another strip/screen/strip

Illustration 3-6—The two layers of screen should be sandwiched between the two screen-mounting strips and all four layers screwed together.

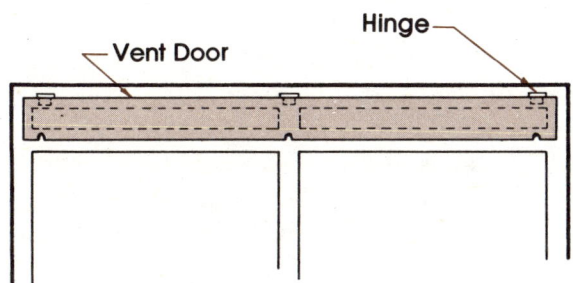

Illustration 3-7—When attaching the vent door, first put the hinges on the door, then attach the door to the frame. Be sure it covers the vent openings equally.

sandwich. The screen does not have to be extremely tight, only taut. There is no weight on the screen; it only needs to be tight enough to hang freely above the baffles. Cut off any extra screen.

When the paint is dry on the glazing frame, attach the vent door you cut and primed in the last chapter and the vent screen. The vent door, 3E, is attached to the glazing frame with three hinges, and held tightly in place with three bolts when closed. The door is hinged, so it can be opened in the summer to ventilate the collector, yet keep rain from getting into the box. Illustration 3-7 shows the position of the door on the glazing frame and the position of the hinges.

First, attach the hinges to the back of the door on the opposite edge of the door from the three notches. Then attach the door to the glazing frame. Be sure the hinges are at the top of the frame, as shown.

With the door attached to the frame, and closed, drill three pilot holes for $3/16$-inch by 2-inch hanger bolts. Each hole should be positioned so the door will close over the hanger bolt, yet be held in place by the cap nut.

To install the hanger bolts, first fasten a cap nut to the end of each bolt. Then, using a wrench, screw the bolt into the pilot hole. Don't fasten the bolt all the way down. Staple the $3/8$-inch weather stripping on the glazing frame in the area shown in illustration 3-8. The weather stripping should fit inside of the hanger bolts, on all sides of the vent opening. Once the weather stripping is applied, close the door and tighten the nuts down until the weather stripping is compressed.

Next, trim a strip of screen 3 inches wide and 46 inches long. Staple this strip over the back of the vent opening. Staple the screen closely enough to be sure insects can't crawl between the screen and the frame. The screen should extend at least $1/2$ inch over the vent opening on all sides.

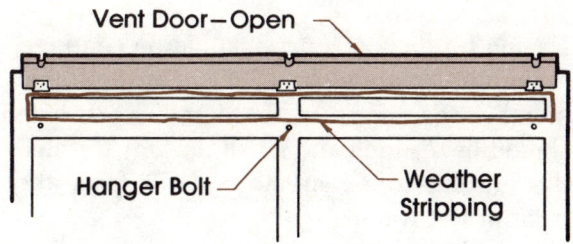

Illustration 3-8—When attaching the weather stripping, be sure it does not interfere with the hanger bolts.

That completes the glazing frame. The next step is to attach the glazing material to the frame.

ATTACHING THE GLAZING

When attaching the glazing to your frame, work carefully and exactly, and you'll have a finished collector without sags or ripples in the glazing. As noted in chapter 1, some parts of the country need only one layer of glazing, while others need two layers. You should have consulted illustration 1-3 (page 10) when ordering your materials to see how many layers of glazing you need. Be sure your glazing material is at room temperature when you start working with it. The recommended materials will stretch and shrink with temperature changes and should be cut and attached at room temperature.

The glazings should be attached to the back of the glazing frame. It is important that the frame be in the exact shape it will be in when installed when the glazing material is attached. The best way to do this is to have the glazing frame attached to the back of the collector box while attaching the glazing material. Rather than just temporarily nailing the glazing frame to the collector box, we have found it best to drill the holes for permanently attaching the glazing frame now, and put small nails through those holes at this time to hold the frame to the back of the collector.

Illustration 3-9 shows the recommended placement of the hanger-bolt holes in the

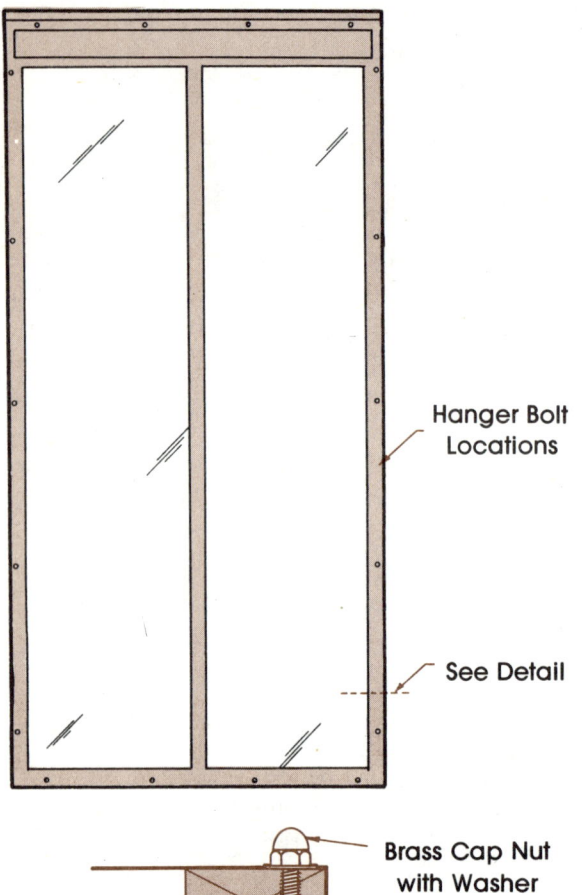

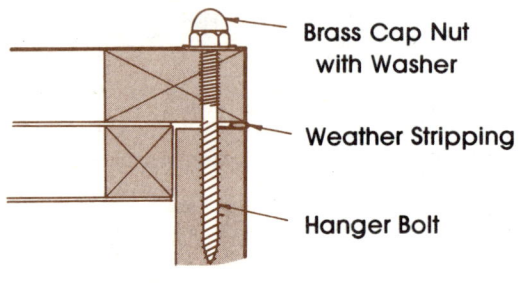

Illustration 3-9—Drill 18 3/16-inch holes through the glazing frame where indicated. Do not drill into the collector box at this time.

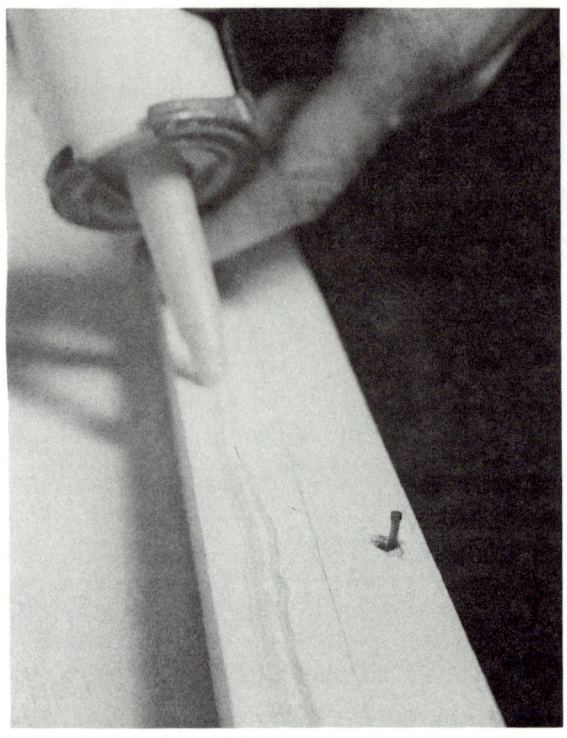

Photo 3-3—First, draw an alignment line, then run a bead of silicone inside the line.

glazing frame. Each hole should be centered 3/8 inch in from the outside edge of the frame member, as shown in the detail. Each hole should be 3/16 inch and go all the way through the glazing frame member. Drill from the front of the frame.

With the holes drilled, position the glazing frame on the BACK of the collector box with the BACK of the frame up, and, using small finishing nails, nail through the holes into the box. Keep the edges of the glazing frame flush with the edges of the collector box. You will have to let the door of the glazing frame overhang one end of the collector box to do this.

Draw a line 5/8 inch in from the inside edge on the back of the side and bottom members of the glazing frame. Also draw a line 1 inch up from the bottom of the center horizontal frame member. These lines will be the alignment lines for the glazing material.

Measure the width and length of the opening marked by the lines. Jot these measurements down on a piece of paper, then remeasure and check them. You need to cut your glazing material to fit the opening marked by the lines exactly. This allows 5/8 inch of glazing for fastening to the frame on all sides.

The glazing materials we recommended can all be cut with a utility knife, heavy-duty scissors, a saber saw with a fine blade, or tin snips. It is best to score the glazing with a sharp instrument where you want to cut it before cutting. With the glazing material cut, test-position it on the glazing frame. If it fits, attach it; if not, trim it so it is within the alignment lines.

Run a bead of silicone caulking about 1/4 inch from the inside edge of each frame member, as shown in photo 3-3, and down the center of the center member.

With someone to help you, position the glazing material on the frame. Be sure you have the glazing facing the right way when you put it on the frame. The recommended glazing materials have an inside and an outside surface; be sure you have the outside surface on the caulking and against the frame members.

First, position the glazing material so one long edge is aligned with the alignment line on one of the side pieces. Then lay the glazing exactly in place, aligned on all other sides with the alignment lines, as sliding it around will smear the silicone. Next, staple it in place every 6 inches. When stapling the glazing material, you almost have to treat it as a piece of fabric. As you staple, have your helper hold the material tight in all directions. Staple first along one long edge, starting in the middle and working toward the ends. Then staple the ends up to the center strip and do the center strip. Then staple the other side and finally the rest of the ends. This should give you a tight-fitting layer of glazing material with no wrinkles.

If you are only putting on one layer of glazing, you are done at this point. If you are using two layers, apply another bead of silicone caulking on the glazing about ½ inch in from the inside edge of the frame members. This will seal the spacers to the first layer of glazing. By sealing the spacers to both layers of glazing, you will prevent condensation from forming between the layers of glazing.

With the caulking on the glazing, position the spacers so they are flush with the inside edge of the frame members, as shown in illustration 3-10. Attach the spacers with 18 1¼-inch #8 wood screws, putting them about every 16 inches. Attach the bottom horizontal spacer first, then the three vertical spacers, and, last, the top horizontal spacer.

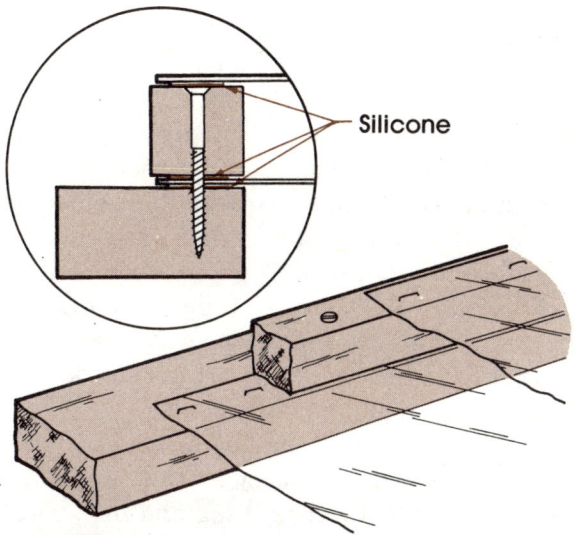

Illustration 3-10—The spacers must be exactly flush with the inside edges of the glazing frame members.

Before attaching the second layer of glazing, clean the inside surface of the attached glazing. This will be your last chance to clean this surface. Measure the distance from the outside edges of the spacers, and cut your second piece of glazing about ⅛ inch smaller on each side. Clean the outside surface of the second glazing before attaching it, and again, be sure you have the outside facing the other layer of glazing. Attach this layer of glazing the same as you did the first layer. The only thing holding this layer of glazing in place will be the silicone caulking and staples. There will be no spacers screwed on top of this layer, so put your staples every 4 inches. Lift the glazing frame off of the back and remove the positioning finishing nails.

That completes attaching the glazing; all that remains is to attach the glazing frame to the collector. The frame is attached with a series of hanger bolts and cap nuts, the same way the vent doors are held closed on the collector box and the glazing frame.

Position the glazing frame on the collector box, making sure it is aligned exactly. Drill ³⁄₃₂-inch pilot holes into the edge of the collector box frame, using the hanger-bolt holes in the glazing frame to position the pilot holes. When you've drilled one hole, screw a hanger bolt into the collector box, and move on to the next hole. Put cap nut on the ends of the hanger bolts, and tighten the bolts into place.

That completes the glazing frame. Turn to the next chapter, and install your hot air heater. The rest of this chapter describes the building of a horizontal glazing frame.

HORIZONTAL GLAZING FRAME

To build a glazing frame for a horizontal hot air heater, use this part of the chapter with blueprint sheet 4 and the horizontal cutting diagram on blueprint sheet 6. The first step in building your glazing frame is to cut the pieces. Then they will be assembled and painted, the glazing material will be attached, and the glazing frame will be attached to the collector box.

Look at blueprint sheet 4 for a few minutes. What you are about to build is a frame to cover the opening of the collector box you have already built. If you built your collector box well, the measurements presented here will work perfectly. If you made any modifications, or had problems, check the overall outside edge measurements of the collector box to be sure the glazing frame will fit. The outside-to-outside measurements of the glazing frame should be 96 inches wide and 47¼ inches high. The frame is designed to fit exactly over the frame of the collector box, but under the overhang of the top. Measure your collector box to be sure your frame will fit.

Begin by cutting the seven frame members. There are four vertical frame members, 4A, and three horizontal frame members, 4B. These are all cut from 1 × 2 stock. Go through your supply of remaining 1 × 2s, and select good sections of the short and long lengths. From the short lengths, cut four vertical frame members, 4A, exactly 47¼ inches long. Be sure both ends of these pieces are square. You will only have a small amount of scrap on these pieces, so carefully trim the ends to be square, without making the pieces too short. Then from your remaining 8-foot lengths of 1 × 2 material, you need three horizontal frame members, 4B, each a full 8-foot length; there should be no scrap for trimming the piece square.

With all seven pieces cut, label them. This may sound like a trivial step, but in reality it is extremely important. If you label your pieces properly, the cutting of the half-lap joints and fitting together of the frame will be very easy—almost foolproof. Pay careful attention to the following instructions.

Lay the seven pieces side by side, three long together and four short together. Select the one best long and two best short pieces; these will be your center pieces. The center members are the most critical, as they do not get bolted to the collector frame and should be the straightest and strongest pieces you have. Check to be sure they are not only knot-free, but straight and square.

With the center members selected, lay them out on a flat surface. Put the long member with its best face up, and the short members with their best face down. This is extremely important as you will be cutting all your joints on the sides that are labeled. With the center members laid out, put the side members next to the center members, again, with the good side of the short members down, and the good side of the long members up.

At this point you should have two piles of wood, one of four short pieces with their good sides facing down, and a second pile of three longer pieces, good side up, all labeled as shown in illustration 3-11.

Now you are ready to mark each piece for the joint areas. The glazing frame is assembled by cutting a half-lap joint at each end of each of these seven pieces, and then cutting a center half-lap in each piece. The center half-lap is not exactly in the center of each piece, as you will see in a few minutes.

First, mark the half-laps on the short, vertical members. Keep the four pieces together, and lay the long, horizontal frame member labeled *top* across one end of all four vertical members, as shown in photo 3-4. Keep the edge of the horizontal member exactly flush with the ends of the four vertical members, and mark the width of the horizontal member across all four vertical members. Label these joint areas just marked *top*.

Return the top horizontal member to its

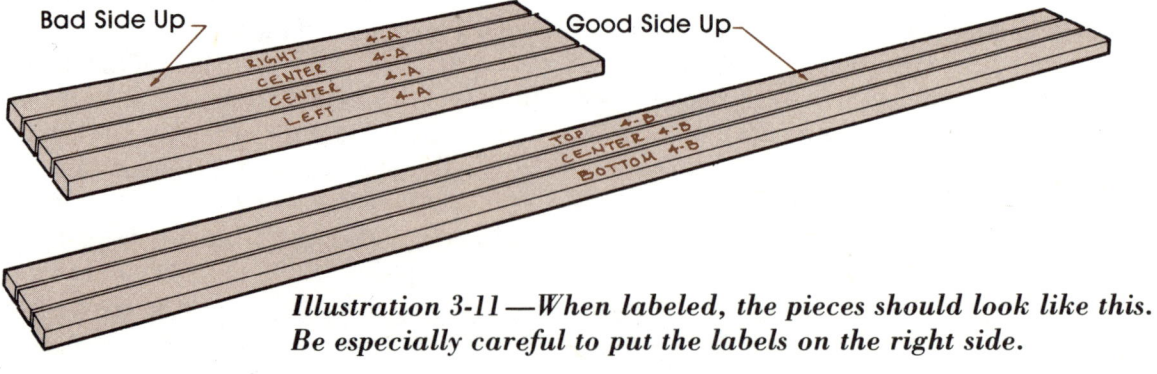

Illustration 3-11—When labeled, the pieces should look like this. Be especially careful to put the labels on the right side.

Photo 3-4—*Keep the edges aligned with the marking piece when marking the joint areas.*

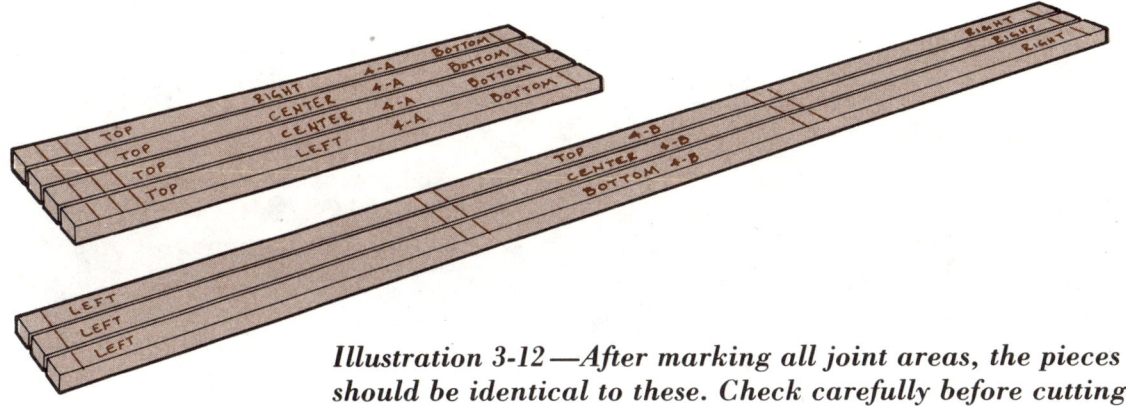

Illustration 3-12—*After marking all joint areas, the pieces should be identical to these. Check carefully before cutting.*

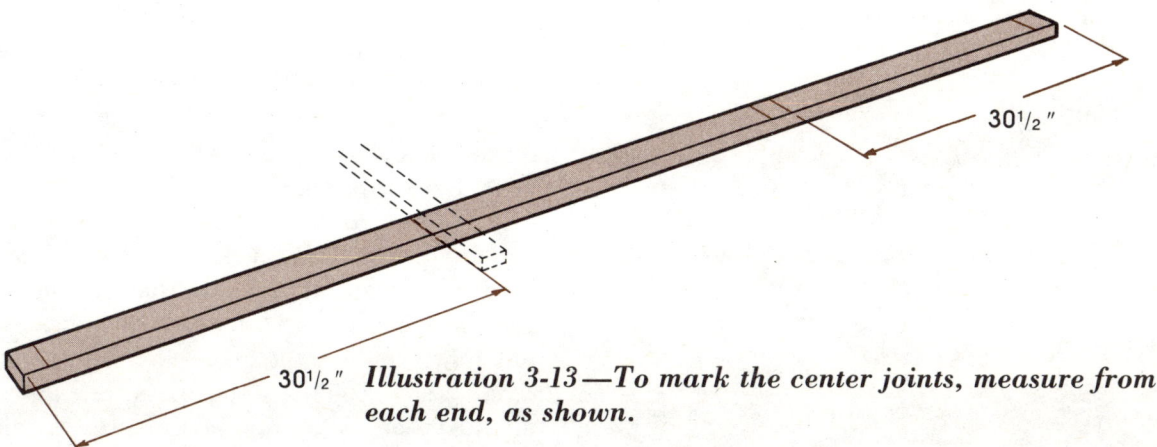

Illustration 3-13—*To mark the center joints, measure from each end, as shown.*

original pile of pieces, select the horizontal piece labeled *bottom*, and do the same marking procedure on the other end of the vertical members. Label this area *bottom*.

To mark the center joint area on the four vertical members, keep them together, and measure 1¾ inches down from the bottom of the top half-lap joint on each piece. Align the top edge of the center horizontal piece with these marks, and mark its width across all the the vertical members. This will require two pencil lines. When finished, the bad side of the vertical members should look like those shown in illustration 3-12.

Return the horizontal members to their original grouping, as shown in illustration 3-11. Take the vertical member labeled *left*, align it with the left ends of the three horizontal pieces, and mark its width as you did on the vertical members. Label this joint area *left* on the good side of the wood. With this done, do the same with the right vertical piece on the other ends of the horizontal members, and label this area *right*.

All that remains is to mark the two center joint areas. Illustration 3-13 shows the measurement for this marking. Repeat the procedure from the other end of the three horizontal pieces to get the positions for the vertical center pieces. Then lay each actual piece on the alignment marks, and mark its width. If you have wood of a width other than 1½ inches, your center frame members will be slightly off-center, but you will never notice it.

Solar Air Heater

That concludes labeling the frame members. Your markings on the seven pieces should be like those in illustration 3-12.

HALF-LAPS

The next step is to cut all the half-lap joints you just marked. All parts of the glazing frame are assembled with half-laps. Using this method, the outside surface of the frame will be perfectly smooth when assembled.

A half-lap joint is made by removing half the thickness of both pieces of wood at the area to be joined. The two pieces are then attached at the joint area with glue and two small screws. This is a strong joint, due to the increased amount of gluing surface and the actual interlocking of the two pieces of wood with screws, as shown in illustration 3-14.

Once you get the technique down and your equipment set up, cut all the joints at one session. Remember to cut all pieces on the labeled side. This will produce a finished frame with the best face of the wood facing out.

There are several ways to cut half-lap joints. If you have a radial arm or table saw, most likely you have made half-lap joints before. All you have to do is set the saw for exactly half the thickness of your wood, and cut within the lines you've marked on the pieces. The only thing to watch out for, once you are set up, is that the saw does not vibrate loose and move to a different setting.

If you don't have a radial arm saw, don't worry; you can make joints that are just as good with a circular saw, or even a handsaw, but it will take a bit longer. If you have a circular saw, you'll cut the joints the same as on a stationary power saw. First, set the depth of your circular saw to half the thickness of your wood. Check your setting by cutting a couple of joints on scrap pieces of wood. Be sure the scrap is the same thickness as the frame members you'll be joining together. Once you have the depth properly set, you can either make many cuts with the saw to remove all the wood, or leave small strips of wood between saw cuts as shown in photo

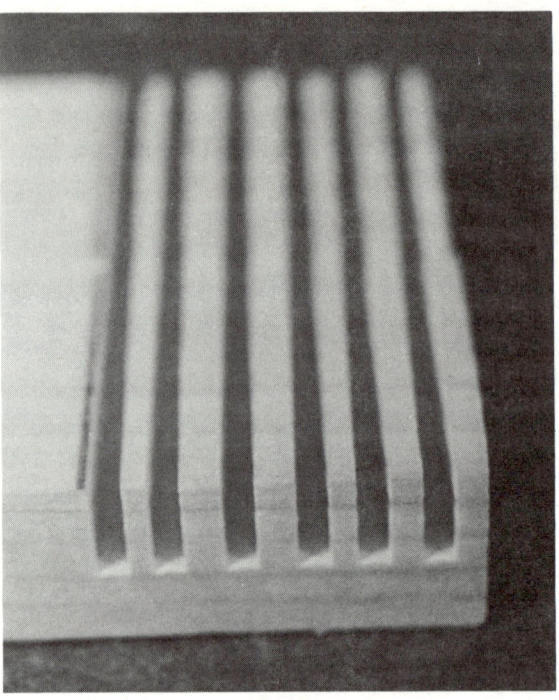

Photo 3-5—To cut a half-lap, set the saw depth and make a number of passes with the saw.

3-5, and chisel them off later. Work slowly and wear safety glasses. It is far better to make the joint too narrow at first and come back and cut out some additional wood, than to cut too much wood the first time and end up with a loose fit.

When checking a joint, test it with the exact piece of wood that will be joined to it. Making good, tight, smooth joints is worth spending a little extra time. Loose joints are susceptible to moisture seeping into the joint, eventually rotting the wood.

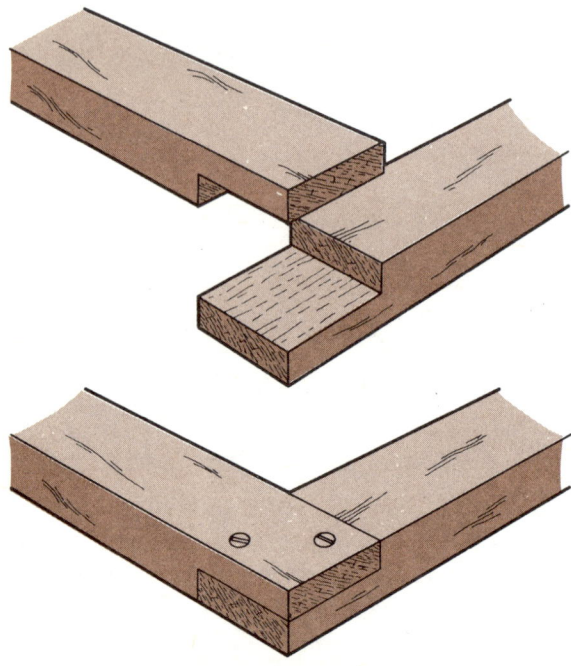

Illustration 3-14—A half-lap joint is easy to cut and very strong. The pieces should be glued and screwed together.

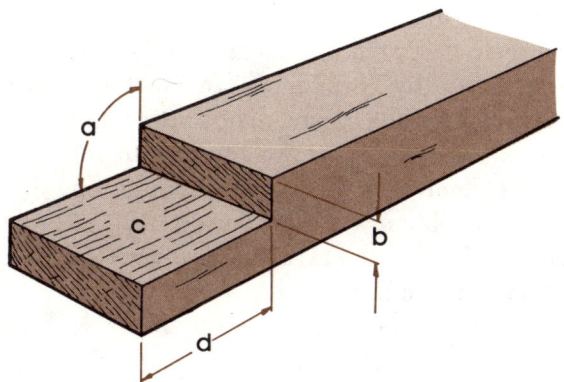

Illustration 3-15—When cutting half-laps, pay attention to four areas: a, the angle of the cut; b, the depth of the cut; c, the levelness of the cut area; and d, the size of the joint area.

If you don't have any power tools and have to use a handsaw to cut the joints, work slowly, removing only a little wood at a time. Keep the area where wood is removed level and uniform as you go. Don't try to remove all of the joint in one cut-and-chisel effort. Work at it slowly, checking often, and you'll end up with joints as good as those made with power tools. Illustration 3-15 shows the areas to be concerned with when cutting half-lap joints. Check each area as you go to be sure you end up with a tight joint.

ASSEMBLING THE FRAME

With all 24 half-laps cut, the next step is to assemble the frame by gluing and screwing each joint. The frame must be assembled all at one time to get it square, yet you have to work quickly to prevent the glue from drying.

The best way to save time during assembly is to predrill two holes at each joint. These holes should only go in the unlabeled side of the horizontal frame members, the long pieces. Each joint will be fastened with two 5/8-inch #6 wood screws. This size screw requires a 9/64-inch hole. Drill the holes in a diagonal position at each joint, as shown in illustration 3-14. Slightly countersink each hole, so the head of the screw will pull flush with the wood when tight. You do not need pilot holes in the vertical frame members, as this size screw will quickly work its way into them without the aid of a pilot hole.

To assemble the frame, lay the two side vertical frame members on a flat surface, with their labels facing up. Then test-position the top and bottom horizontal frame members between the two vertical members.

Once you have the four pieces properly positioned and are sure they fit together well, apply glue to both faces of the four corner joints, and screw the frame members together. The best way to make your frame square is to use the back of the already assembled collector box as your work surface. The glazing frame and the collector box should align exactly. If you work on the back of the collector, once all the pieces align with the plywood, tighten both screws at each joint.

If you don't work on the back of the collector box, put only one screw in each joint, and make them just snug, not tight. Then measure the distance diagonally from corner to corner to see if the frame is square. If the measurements are the same, you've got a square frame; if not, you've got to adjust it. Use your combination square to check each corner until it is square, and then remeasure it for overall squareness. The frame must be in square before you tighten the screws. With the screws tight, remeasure for square, and put the second screw in each joint.

Once you have the four outside pieces of the frame correctly assembled, glue and screw the other frame members in place. At this point, your frame should be firmly assembled. Let it set for a few hours for the glue to dry.

Next, if you are using two layers of glazing, cut the frame spacers. These are six 3/4-inch by 3/4-inch pieces of wood used to keep the two layers of glazing apart. As noted earlier, baluster stock is the ideal size, but if you can't buy that, you can either rip the strips from a 1 × 6 or have it done at the lumberyard. You need four vertical spacers, each measuring 41 inches, and two horizontal spacers, each measuring 94 3/8 inches.

With the frame assembled and the spacers cut, prime and paint all pieces. If you did not prime the vent door, 4E, in the last chapter, do so now. As noted earlier, use a primer that is formulated to go with the exterior paint you will use.

The glazing frame and door can be painted any color you want, although theoretically, white is the best color. White will reflect some sunlight into the heater, while a darker color would absorb the light, reducing the effectiveness of the heater slightly.

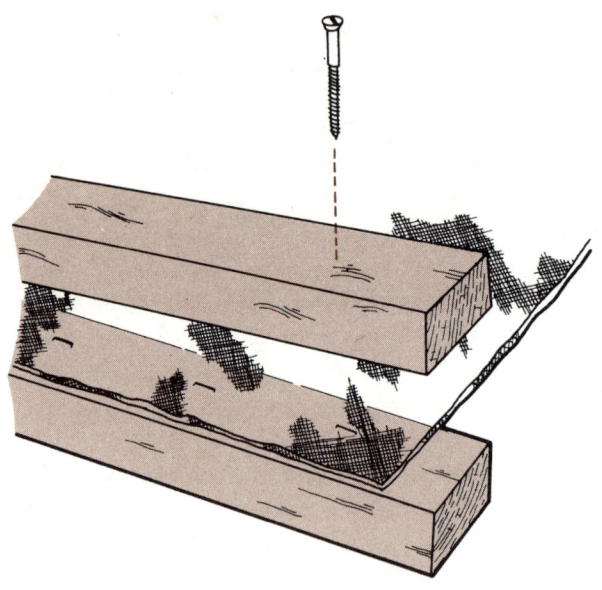

Illustration 3-16—The collector screens should be sandwiched between two screen-mounting strips, as shown.

Before painting, be sure all knots, cracks, and spaces at the joints are caulked with an outdoor wood filler. You want the finished frame to be as smooth as you can get it. Don't forget to fill the screw holes in the back of the frame.

While the primer and paint are drying, attach the collector screen to the screen-mounting strips. The screen serves to increase the amount of collector surface inside the collector box without greatly inhibiting airflow. If you can find it, GET BLACK ALUMINUM WINDOW SCREEN. DO NOT GET FIBERGLASS SCREEN; it will not collect heat. If you cannot find black screen, get regular aluminum insect screen, and paint it black. Painting screen is a real time-consuming chore, so check around for black aluminum screen; see if someone can order it for you.

The collector uses two pieces of screen, each approximately 45 inches by 96 inches. The extra screen should be stapled over the vent openings on the inside of the box. Staple the screen well, so insects can't crawl under the screen and into the box. The two 96-inch-long layers are attached between two screen-mounting strips, 4F, at each end. Lay out the two pieces of screen, one on top of the other, and staple them to one of the screen-mounting strips with 3/8-inch staples. Then screw another screen-mounting strip to the first with five 1¼-inch #8 wood screws, sandwiching the two layers of screen between two screen-mounting strips, as shown in illustration 3-16.

With one end of the collector assembled, put it into a pair of screen-mounting brackets in the collector box. At the other end of the collector box, put another screen-mounting strip in the brackets. Stretch the screen as tight as you can from one end to the other, and staple the screen to the mounting strip. Then screw the last mounting strip to the other, making another strip/screen/strip sandwich. The screen does not have to be extremely tight, just taut. There is no weight on the screen; it only needs to be tight enough to hang freely above the baffles. Cut off the excess screen.

When the paint is dry on the glazing frame, attach the vent door, 4E, you cut and primed in the last chapter, and the vent screen. The vent door is attached to the glazing frame by four hinges and fastened with four bolts when closed. The door is hinged so it can be opened in the summer to ventilate the collector, yet keep rain from getting into the box. Illustration 3-17 shows the position of the door on the glazing frame and the position of the hinges.

First, attach the hinges to the back of the door on the opposite edge of the door from the notches. Then attach the door to the glazing frame. Be sure the hinges are at the top of the frame, as shown.

With the door attached to the frame, and closed, drill four pilot holes for $3/16$-inch by 2-inch hanger bolts. The pilot hole position should be marked with the door closed. Each hole should be positioned so the door will close over the hanger bolt, yet be held shut by the cap nut.

To install the hanger bolts, first fasten a cap nut to the end of each bolt. Then, using a wrench, screw the bolt into the pilot hole. Don't fasten the bolt all the way down. Staple the 3/8-inch weather stripping on the glazing frame in the area shown in illustration 3-18. The weather stripping should fit inside of the hanger bolts on all sides of the vent opening. Once the weather stripping is applied, close the door and tighten the nuts down until the weather stripping is compressed.

Next, trim a strip of screen 3 inches wide and 94 inches long. Staple this strip over the back of the vent opening. Staple the screen closely enough to be sure insects can't crawl between the screen and the frame. The screen should extend about ½ inch over the vent opening on all sides.

That completes the glazing frame. The next step is to attach the glazing material to the frame.

ATTACHING THE GLAZING

When attaching the glazing to your frame, work carefully and exactly, and you'll have a finished collector without sags or ripples in the glazing. As noted in chapter 1, in some parts of the country you will need only one layer of glazing, while in others you will need two layers. You should have consulted illustration 1-3 (page 10) when ordering your materials to see how many layers of glazing you need. Be sure your glazing material is at room temperature when you start working with it. The recommended materials will stretch and shrink with temperature changes and should be cut and attached at room temperature.

The glazings should be attached to the back of the glazing frame. It is important that the frame be in the exact shape it will be in when installed when the glazing material is attached. The best way to do this is to have the glazing frame attached to the back of the collector box while attaching the glazing material. Rather than just temporarily nailing the glazing frame to the collector box, we have found it best to drill the holes for permanently attaching the glazing frame now, and put small nails through those holes at this time to hold the frame to the back of the collector.

Illustration 3-19 shows the placement of the hanger-bolt holes in the glazing frame. Each hole should be centered ⅜ inch in from the outside edge of the frame member, as shown in the detail. Each hole should be ³⁄₁₆ inch and go all the way through the glazing frame member. Drill from the front of the frame.

With the holes drilled, position the glazing frame on the BACK of the collector box with the BACK of the frame up, and, using small finishing nails, nail through the holes into the box. Keep the edges of the glazing frame flush with the edges of the collector box. You will have to let the door of the glazing frame overhang one side of the collector box to do this.

Draw a line ⅝ inch in from the inside edge of the side and bottom members of the glazing frame. Also draw a line 1 inch up from the bottom of the center horizontal frame member. These lines will be the alignment lines for the glazing material.

Measure the width and length of the opening marked by the lines. Jot these measurements down on a piece of paper, then

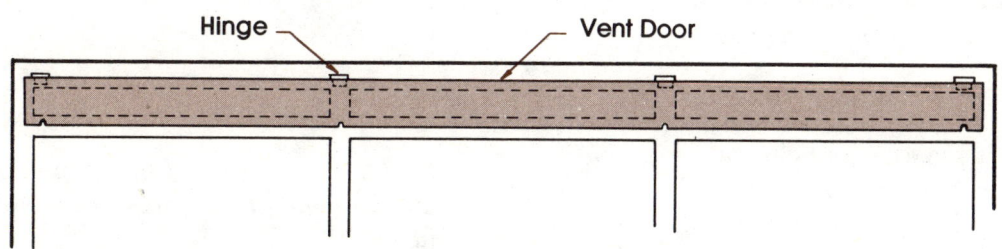

Illustration 3-17—The vent door is attached to the glazing frame with four hinges. Be sure the door overlaps the vent openings equally.

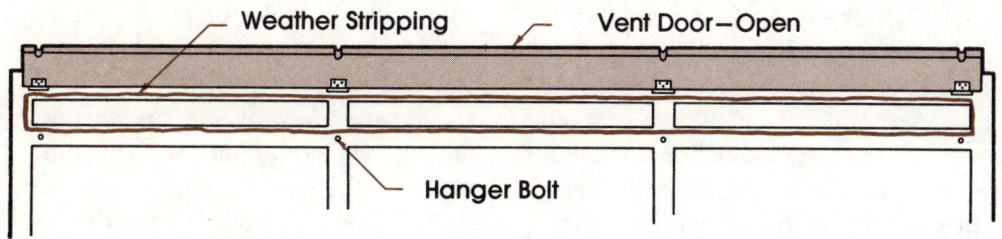

Illustration 3-18—Be sure the weather stripping does not interfere with the hanger-bolt placement.

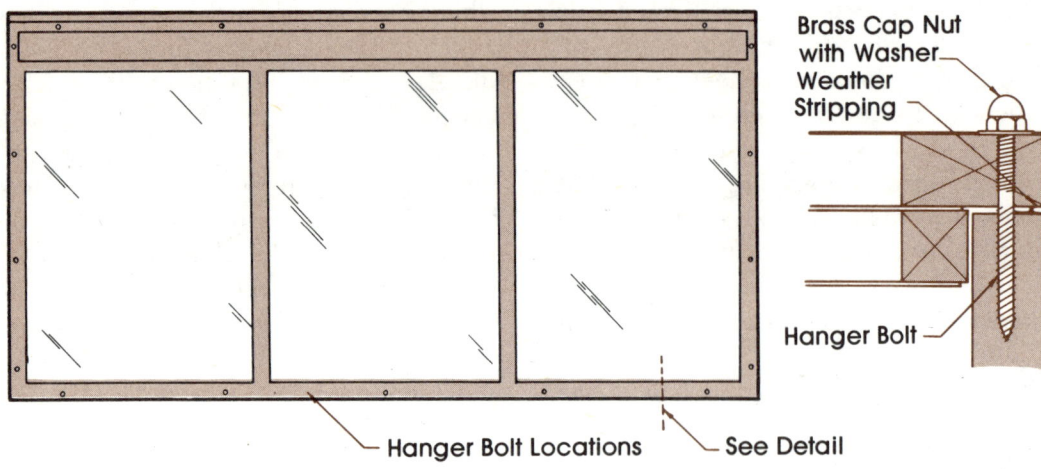

Illustration 3-19—Hanger-bolt placement is critical for a well-sealed collector. Don't drill the shank holes into the collector box, just through the glazing frame.

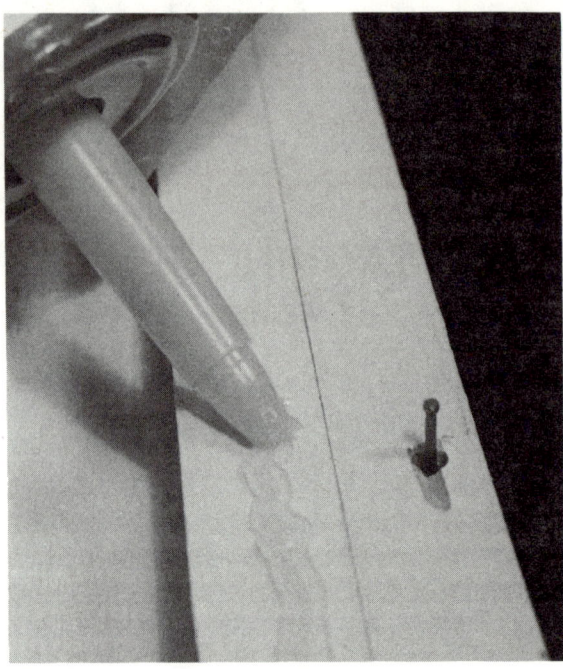

Photo 3-6—Run a bead of silicone along the inside edge of the glazing frame members.

remeasure and check them. You need to cut your glazing material to fit the opening marked by the lines exactly. This allows about 5/8 inch of glazing for fastening to the frame on all sides.

The glazing materials we recommended can all be cut with a utility knife, heavy-duty scissors, a saber saw with a fine blade, or tin snips. It is best to score the glazing with a sharp instrument where you want to cut it before cutting. With the glazing material cut, test-position it on the glazing frame. If it fits, attach it; if not, trim it so it is within the alignment lines.

Run a bead of silicone caulking about 1/4 inch from the inside edge of each frame member, as shown in photo 3-6, and down the center of the two center members.

With someone to help you, position the glazing material on the frame. Be sure you have the glazing facing the right way when you put it on the frame. The recommended glazing materials have an inside and an outside surface; be sure you have the outside surface on the caulking and against the frame members.

First, position the glazing material so one long edge is aligned with the alignment line on one of the long horizontal pieces. Then lay the glazing exactly in place, aligned on all other sides with the alignment lines, as sliding it around will smear the silicone. Next, staple it in place every 6 inches. When stapling the glazing material, you almost have to treat it as a piece of fabric. As you staple, have your helper hold the material tight in all directions. Staple first along one long edge, starting in the middle and working toward the ends. Then staple the ends and the center strips. Then staple the other long edge. This should give you a tight-fitting layer of glazing material with no wrinkles.

If you are only putting on one layer of glazing, you are done at this point. If not, apply another bead of silicone caulking on the glazing, about 1/2 inch in from the inside edge of the frame members. This will seal the spacers to the first layer of glazing. By sealing the spacers to both layers of glazing, you will prevent condensation from forming between the two layers of glazing.

With the caulking on the glazing, position the spacers so they are flush with the inside edge of the frame members, as shown in illustration 3-20. Attach the spacers with 18 1¼-inch #8 wood screws, putting them about every 16 inches. Attach the bottom horizontal spacers first, then the four vertical spacers, and the top horizontal spacer last.

Before attaching the next layer of glazing, clean the attached glazing. Measure the distance from the outside edges of the spacers, and cut your second piece of glazing about ⅛ inch smaller on each side. Clean the second glazing before attaching it, and again, be sure you have the outside surface facing the other layer of glazing. Attach this layer of glazing the same as you did the first layer. The only thing holding this layer of glazing in place will be the silicone caulking and staples. There will be no spacers screwed on top of this layer, so put your staples every 4 inches. Lift the glazing frame off of the back and remove the positioning finishing nails.

That completes attaching the glazing; all that remains is to attach the glazing frame to the collector. The frame is attached with a series of hanger bolts and cap nuts, the same way the doors are held closed on the collector box and the glazing frame.

Position the glazing frame on the collector box, making sure it is aligned exactly. Drill $3/32$-inch pilot holes into the edge of the collector box frame, using the hanger-bolt holes in the glazing frame to position the pilot holes. When you've drilled one hole, screw a hanger bolt into the collector box, and move on to the next hole. Put cap nuts on the end of the hanger bolts and tighten the bolts into place.

That completes the glazing frame. Turn to the next chapter, and install your hot air heater.

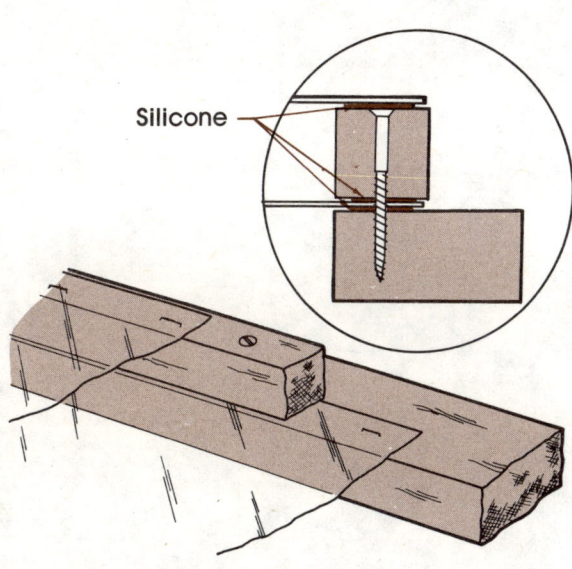

Illustration 3-20—The spacers must be exactly in line with the inside edge of the glazing frame members.

4 Installing the Heater

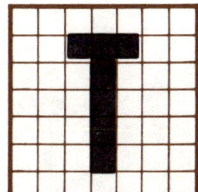

To some extent, this chapter differs from the other chapters in this book. Up to now, we've given you very exact instructions for building the heater. However, from this point on, we can only give guidelines and suggestions. We'll tell you what to look out for, and what you have to do, but we can't tell you how to do it; that depends on the type of wall to which you'll be fastening your collector.

We give directions for two types of walls, a solid masonry wall and a wood frame wall. There are hundreds of variations of each, so we will be giving you general rules of thumb for installation, and you'll be on your own to figure out the exact details. If you don't feel comfortable cutting holes in the side of your house after reading this chapter, by all means get someone to help you. We have installed heaters on several different wall surfaces and have found it takes two people anywhere from three to five hours to put up the unit. There will be some added time touching up and finalizing details, but, basically, figure an afternoon to mount the unit on your wall with two people working, and another afternoon or day to get everything else done with only one person working. If you are working on the second story of your home, allow more time; you'll be amazed at how many tools you leave on the ground and how many trips up and down you'll be making.

We did not include any tools for installing the unit in the Tools section, nor did we include any materials for installing the unit in the Materials section. These will vary, depending on the type of installation you have, so go through this chapter and estimate what you will need.

One word of caution before we start: You will be breaking the thermal envelope of your home when installing the collector. If you don't install it properly and weather-strip and caulk the unit tightly, you will have air infiltration around the collector and into your home. This infiltration of freezing cold air can deplete the energy gains of the collector quickly. We will point out all areas of concern as we go through the instructions; pay close attention to these details, do your weatherproofing job well, and you'll end up with an installation that will last and provide comfort for many years.

Before you start installing your heater, you have to find out what type of walls you have on your house, if you don't already know. This will call for some detective work. There are several ways you can find out. One way is to remove an electric recepticle along an exterior wall, peak behind the recepticle with a flashlight, and see what is there. Turn off the electricity before doing this. Another way is to look around window openings, doors, and other interruptions in the wall. You can measure the thickness of the wall to get an idea of what you will be dealing with. Cement-block walls are normally 8 inches thick, with added space for exterior stucco, interior plaster, or interior furring strips and plaster or wallboard. Brick walls are normally at least two layers of brick thick; older homes are three layers. Many new homes will be frame, covered with a brick veneer. In this

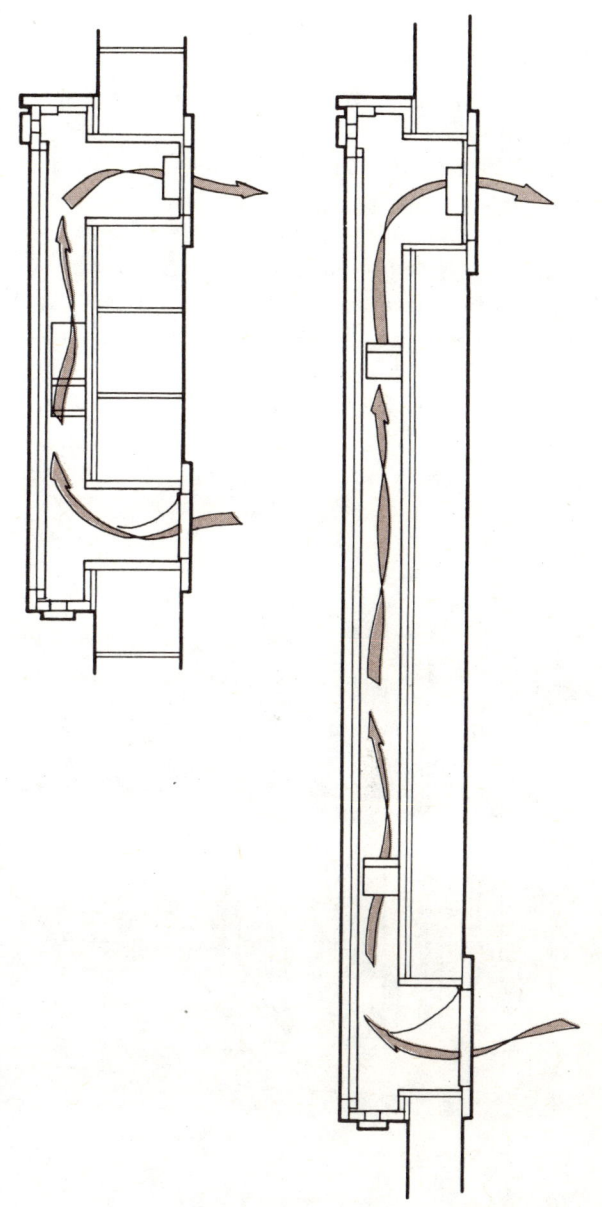

Solar Air Heater

case, you will have to proceed on the inside as for a standard frame wall, and remove bricks where you have to on the outside as for a masonry wall.

When installing your collector, you can proceed in one of two ways after deciding which is the most critical point: where the vents are positioned inside your room, or where the collector is positioned on the outside of your house. Depending on this decision, you will work from the inside to the outside or the reverse. For now, find out what type of wall you have, and follow the directions for that type of installation.

We'll give directions first for a frame wall, then for a masonry wall. The instructions in each section deal only with attaching the collector to the wall. The very last part of the chapter explains the building of the duct work and vent grilles. Use the appropriate part of the book for your type of wall for mounting the collector, then turn to page 61 to see how to finish your installation. The finishing details are the same for both a vertical and a horizontal heater and for either type of wall.

FRAME WALL

This is the easier type of installation. A frame wall is shown in illustration 4-1. For this type of installation, we think it is better to start on the inside and position the vents exactly where you want them, then move to the outside of the house and position the collector over the vents. In some cases you may want to position the collector outside first to maintain the exterior aesthetics of your home, and then position the vents inside. This is fine; just work backward from the instructions in this part of the chapter.

For frame walls, your vent openings should extend the full width of the space between frame members. Normally studs are 16 inches on center, giving a vent width of about 14 ½ inches. If your studs are 24 inches on center, this is fine. We recommend what we call "the rat-hole method" of locating wall studs. This requires that you mark the center point of your ideal placement of the bottom

Illustration 4-1—A wooden frame wall normally is made up of vertical frame members 16 or 24 inches on center.

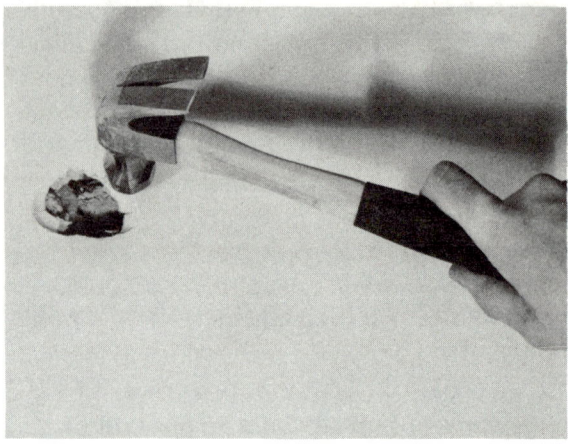

Photo 4-1—Begin the "rat-hole method" by breaking out a small hole in the wall. This hole should be at your ideal spot for the center of a vent.

vent. Be sure you have at least 3 feet of clearance on each side of the center point on the outside wall of the house to mount the collector. We say 3 feet on each side, in case the point you pick as your center point is not in the center of the stud opening, and the stud openings have to be shifted left or right. In a pinch, the vents do not have to be centered inside the collector, but it is better if they are. To use the rat-hole method, start off with a small hole, and carefully enlarge it from side to side until you have located both side studs. Then measure for the height of the vent, mark the top and bottom, and remove the remaining wall covering. Photos 4-1 and 4-2 show this being done. When expanding the opening, work very carefully, checking to be sure there are no wires or pipes; if there are, you will either have to move them or reposi-

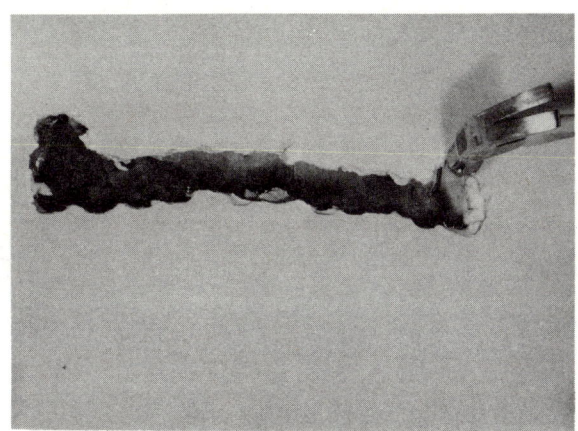

Photo 4-2—*Extend the hole from side to side until both studs are located.*

tion your collector. Don't open up a stud area below or above an electric recepticle; there are sure to be wires there. If you are installing the heater on the ground floor, go into the basement and check to see if you see any pipes in the area where you will be installing the collector.

The vent opening will normally be about 14½ inches wide on a standard stud wall. For an opening of this width, the vent should be from 8 to 10 inches high. On a wall with studs every 24 inches, a vent height of 6 inches is the minimum. A vent opening of about 100 to 150 square inches or more is the goal. If you will be using your heater in a passive mode, without fans, the bigger the vent opening, the better. You may want to consider two openings side by side.

With the bottom vent located, the top vent should align within the same stud cavity for a vertical heater. Try to have the vents as far apart as you can, but stay within the measurements in illustration 4-2 as maximums. These maximum measurements will use the entire collector space, giving you the utmost efficiency. If you exceed these measurements, you'll end up with vents that do not fit within the overall collector opening.

If you cannot get the full measurements, your vents will have to be closer together. Your heater will still work; it just won't be quite as effective. The guideline for vent placement on the wall is to have the bottom vent as close to the floor as you can get it, while still leaving room for the vent trim. A vertical installation in homes with standard 8-foot ceilings will place one vent just below the ceiling, the other just above the floor.

With a horizontal installation, you have a lot more leeway in your placement. Look at illustration 4-3 for maximum vent placements on a horizontal collector. You have to be careful not to exceed the vertical spacing limitation of the vents on a horizontal unit. Pay careful attention to the illustration to avoid this. Illustration 4-3 shows typical 16-inch stud spacing for vent placement in a horizontal unit. Use it as a guide to position your vents properly.

With both interior vents cut, you are ready to mark the positioning of the collector on the outside surface of the wall. The easiest way to do this is to drill a hole at each corner of each vent from the inside through the outside face of the wall. This will mark the vent openings on the outside. From that point, you can move outside.

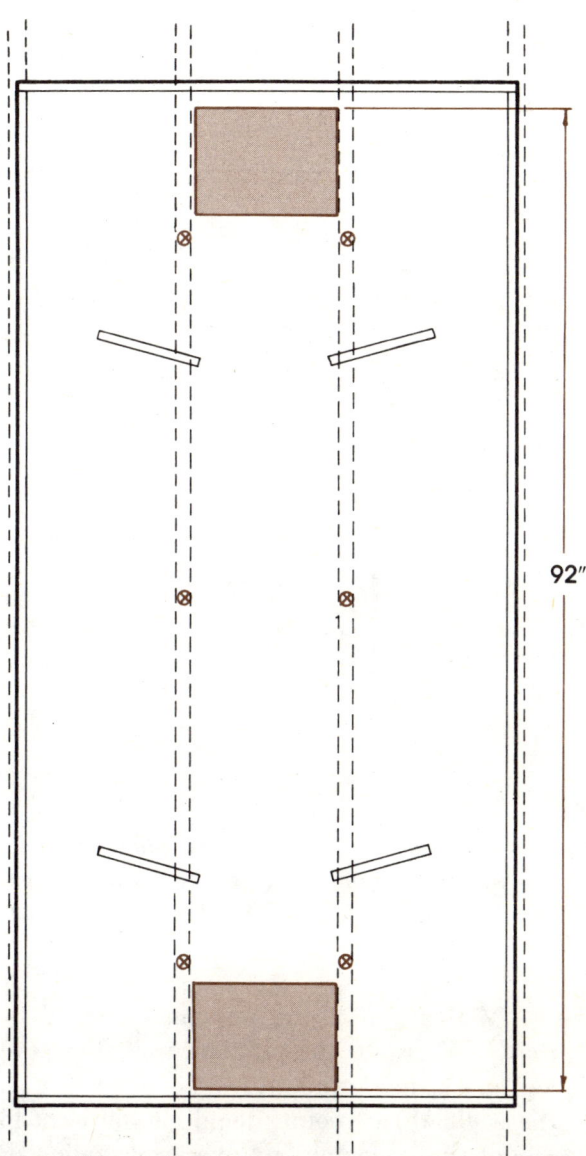

Illustration 4-2—*Maximum vent spacing for a vertical heater on a wall with studs 16 inches on center. This spacing would be the same for a wall with studs 24 inches on center. The Xs indicate where to fasten the collector to the studs.*

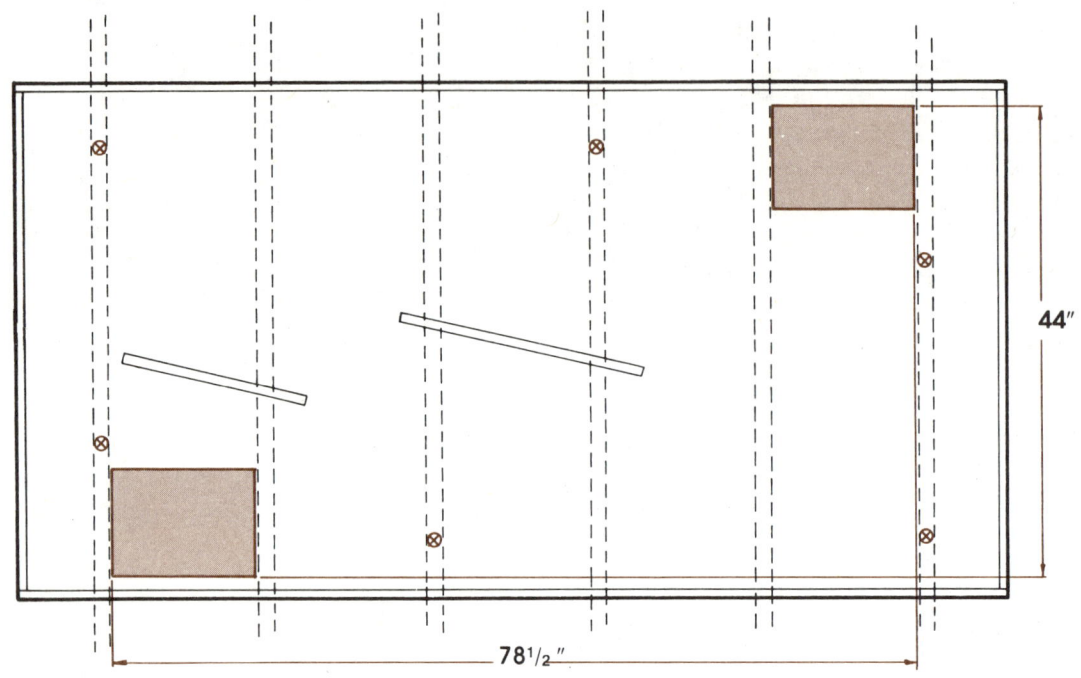

Illustration 4-3 — Maximum vent spacings for a horizontal heater on a wall with studs 16 inches on center. These spacings would change for a wall with studs 24 inches on center. The Xs indicate where to fasten the collector to the studs.

With the two vent openings marked by the holes drilled through the wall, measure and mark the outline of the collector box on the wall. At this point, decide if you want to mount the collector box over your siding or remove the siding and mount the box flat against the sheathing. We think removing the siding is worth the extra work. Remember, your collector will be a permanent part of your home, so why not mount it the best way possible?

To remove the siding, mark the outside measurement of the collector box on the siding, using the vent holes for exact positioning. Use a circular saw with a special blade to cut away the siding. When measuring for the opening to be cut into your siding, allow an extra 1/8 inch on all sides. Only cut deep enough to remove the siding. You want to leave any underlayment, sheathing, or weatherboard in place. With the siding removed, cut out the covering over the vent openings. If your house is insulated, remove the insulation from the vent opening area. If you have a loose, pouring type of insulation, you'll have to slip a temporary stop across the top of the vent opening to prevent the insulation from pouring out of the opening. Work very carefully while doing this, and be sure to wear safety glasses. Photo 4-3 shows aluminum siding cut away and capped with J channel. This gives a factory-finished look to your collector installation.

If you do not want to remove the siding, the collector can be mounted right over the siding. This requires more care when positioning the collector and extra weather stripping and caulking, but it can be done. At this time, just cut out the two vent openings. A saber saw is normally best for this, using the four corner holes as starting points for each vent opening.

With the opening cut away, it becomes a simple matter of locating bolt placement over the studs you have already located. Illustrations 4-2 and 4-3 show the ideal bolt placement for vertical and horizontal collectors on a 16-inch stud wall. The collector is bolted to the studs with lag screws and washers. First, remove the glazing frame from the collector, then drill shank holes through the back of the collector, hold it in place on the wall, and mark the locations of pilot holes on the studs. You will have to crush the back insulation in the collector in the bolt area, or cut it out. The washers should be fastened tight against the plywood back of the collector. With pilot holes drilled in the studs, you may find it best to mark and drill only one or two pilot holes,

Photo 4-3—Cut away aluminum siding as shown. Use J channel to cover the edge, as shown in the detail on right.

fasten the collector in place, and then mark the rest, as holding the collector in place on the wall for any length of time is difficult. Fasten the collector in place on the wall, but don't tighten the screws all the way—just make it snug against the wall. Then go inside and mark the actual vent openings on the back of the collector box.

Then remove the box from the wall, and cut out the vent openings. When you have the vent openings removed, weather-strip the

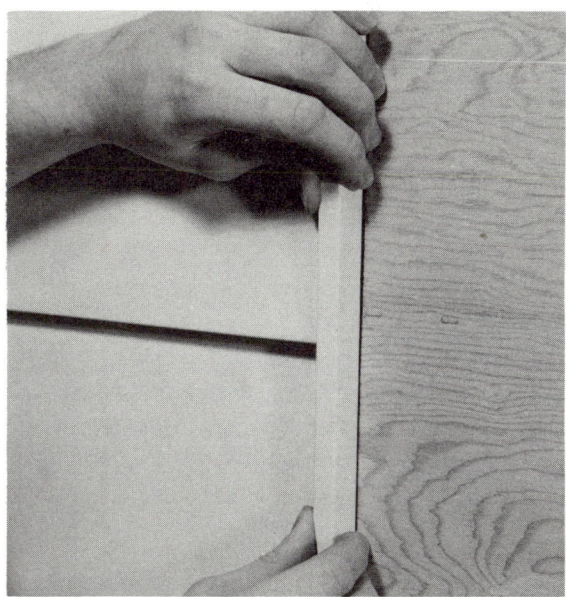

back of the box with closed-cell foam weather stripping. We have found ½-inch-thick by ¾-inch-wide weather stripping to work best. This weather stripping should go around the entire outside perimeter of the back of the box, and around each vent opening. With the box completely weather-stripped, you are ready to install it permanently. Be careful not to disturb the weather stripping when mounting the box.

Bolt the box firmly to the wall. Then go around the entire outside of the box and caulk the gap where it meets the house. Be sure to caulk the top especially well, as you do not want moisure migrating behind the box. At this point you are ready to wire the box and build the vents. These instructions are explained starting on page 61. Move to that section to continue your installation.

MASONRY WALL

If your home has solid masonry walls, you have slightly more work, but don't worry—in some ways they are easier to work with than frame walls. The two biggest problems with masonry walls are positioning the vent openings properly and attaching the collector to the wall. Breaking the masonry out can also be hard work.

The instructions presented here are for brick or cement-block homes. If you have adobe, stone, or some other form of solid masonry, you'll have to wing it somewhat. This chapter will give you general guidelines, but you'll have to figure out your own approach. The goals are the same; knock two vent openings through the walls, and firmly attach the collector to the outside of the wall.

Working with brick or cement block, the key to an easy job is to let the masonry dictate where the collector will be positioned. You can pick the general location, but let the mortar joint dictate the exact vent positioning. Illustrations 4-4 and 4-5 show typical grids for brick and cement-block homes. The best areas to put your vents are marked in color. These positions will give you the most efficient vent positioning for the least amount of work.

There is no exact vent size for the heater; we let the size of the masonry opening dictate the vent opening size. There are acceptable minimums, and we do not recommend you go below those sizes shown on illustrations 4-4 and 4-5.

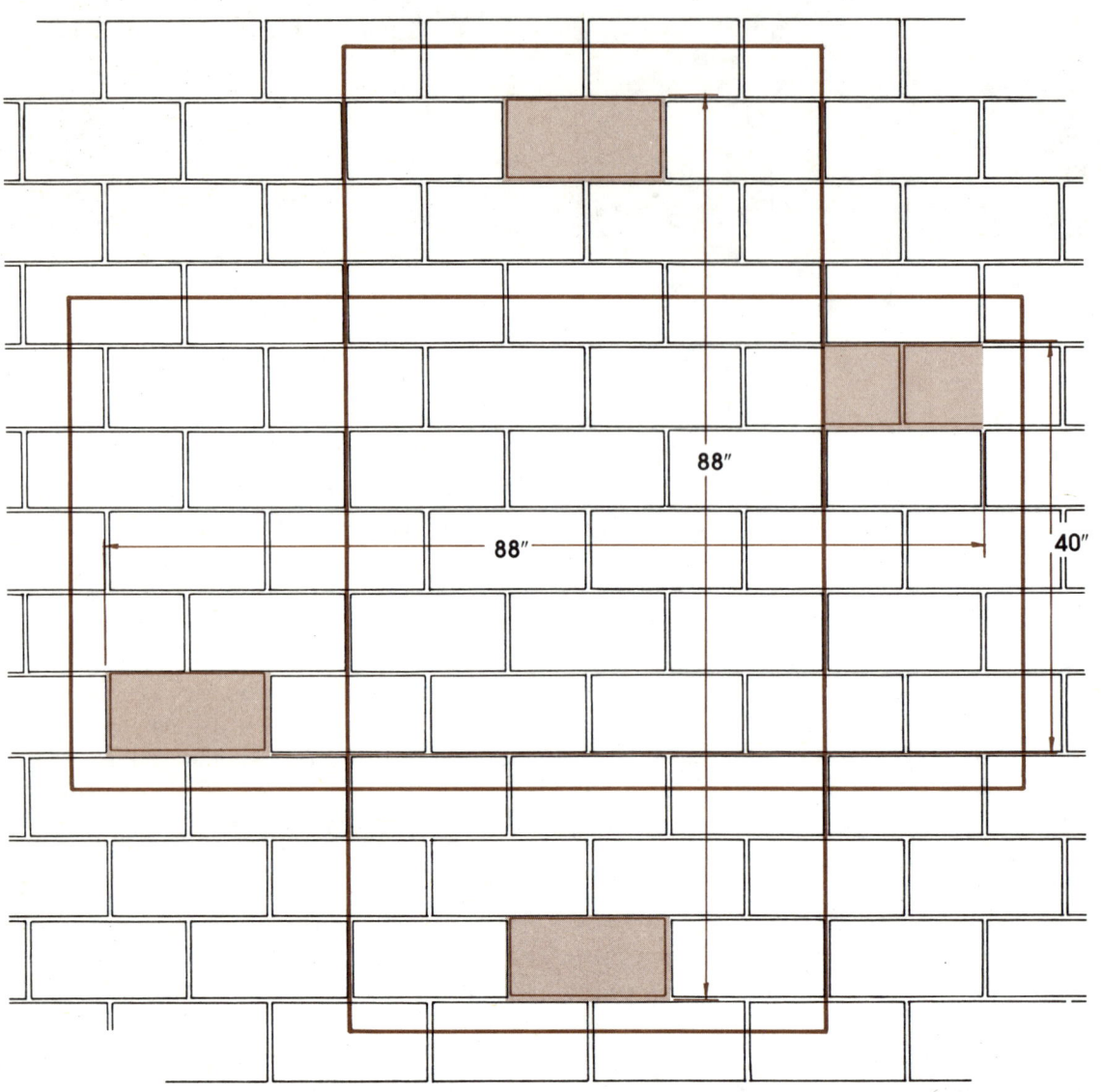

Illustration 4-4—Masonry grid for a cement-block wall with blocks 8 × 16 inches. The outlines show both vertical and horizontal collector placements. The dimensions indicate the maximum spaces between vents.

To start your installation, find a mortar joint in your masonry. On a brick wall or an uncovered cement-block wall this is easy; just look. But on a stucco wall, you have to use what we call "the rat-hole method." This entails marking the ideal vent placement area and slowly chipping away at the stucco in one direction until you find a mortar joint.

Before starting to chip away the stucco, go inside and see if the vent positioning will be all right. You can do this by measuring from a window or door opening or a corner of the wall. You won't be able to position the final vent placement inside exactly, but you should get a good idea of where it will fall. You have to be sure that by following the grid work in illustrations 4-4 and 4-5 your second vent will also fall where you want it to be. To do this, estimate the first vent position, and then measure to estimate where your second vent will fall on the inside of your room. The measurements given on illustrations 4-4 and 4-5 are the maximum spacings for the vents for both brick and cement-block walls and for both vertical and horizontal collectors. These measurements cannot be exceeded, or the vents will be larger than the collector box and you'll have problems. Measure carefully before starting to knock holes in the side of your house. Enough said?

When you find one mortar joint, begin chipping in the perpendicular direction until you find a second joint, then work in the other two directions to chip away the area you will need. Photos 4-4 and 4-5 show the rat-hole method in operation on a cement-block wall with stucco over the block. Note that we started chipping, then expanded the hole until we found one joint, then we worked in the perpendicular direction to find another joint, then worked to find the other two joints.

With one vent opening located, use a square and level or plumb line to lay out the overall positioning of the collector on the wall.

With the perimeter of the collector marked on the wall, again use the rat-hole method for marking out the second vent opening. If you have a cement-block wall and a horizontal collector, you'll have to break a block; if you have brick walls, you'll have to break several bricks. Luckily, the area where you have to break the blocks is not exact, and if the block does not break exactly where you want it, that is OK; close is acceptable.

When working with masonry, always wear safety glasses and leather gloves. Use a flat-blade chisel for chipping away stucco, but use a serrated-edge chisel for cutting blocks or bricks.

When removing cement blocks or bricks, begin by chiseling out the mortar around the unit to be removed. If cutting a unit, remove the mortar only around the part you want to remove. With the mortar removed, begin by chipping away one corner, and move toward the center of the unit, removing the chipped-away parts as you go.

Don't be intimidated about working with masonry. You will have good, square sides for your vents once it is removed. Follow the grids carefully, so that you don't end up with vents spaced too far apart to be covered with the collector box. To avoid this, be sure to

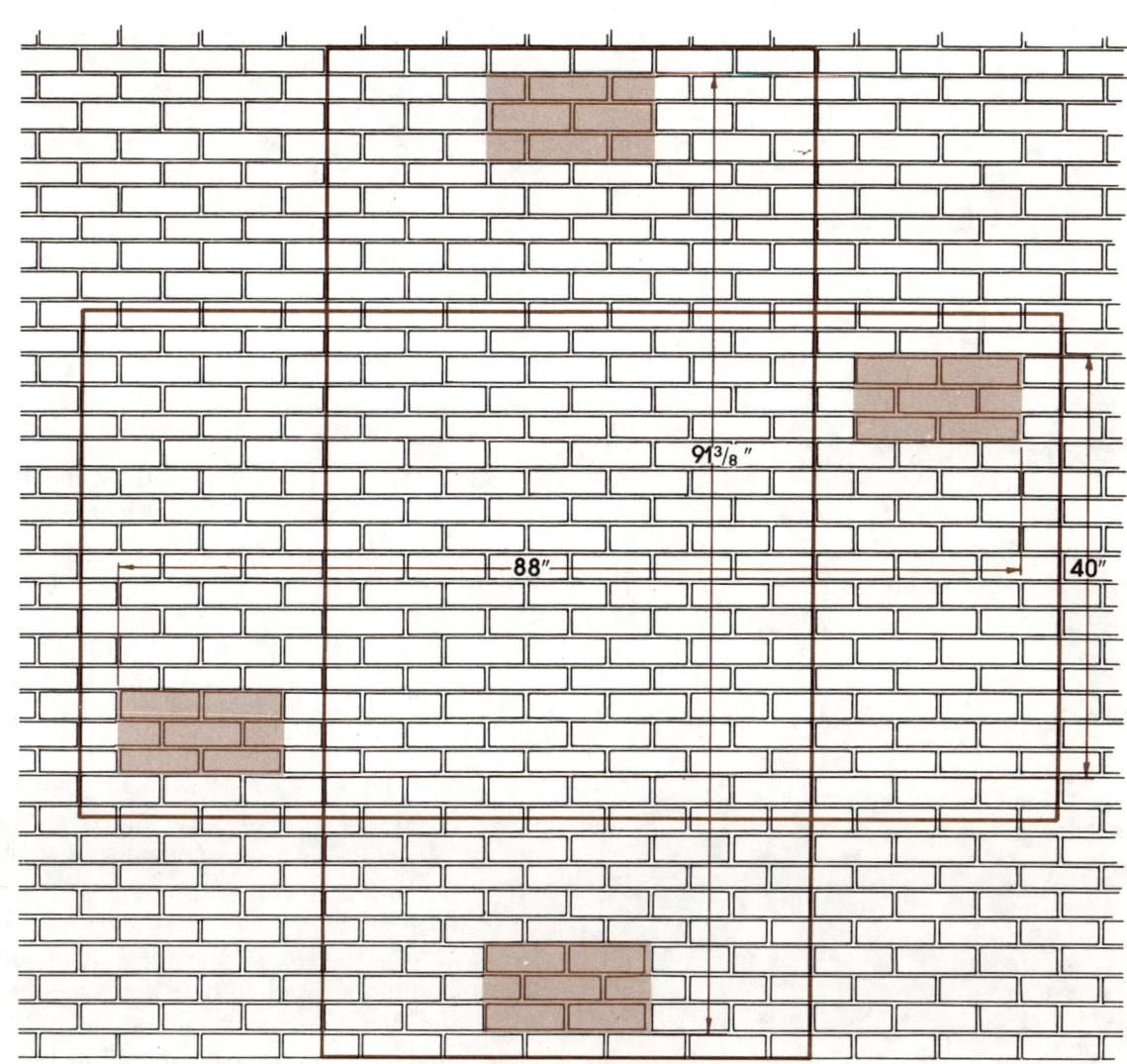

Illustration 4-5—Masonry grid for a brick wall with bricks approximately 2½ × 8 inches. The outlines show both vertical and horizontal collector placements. The dimensions indicate the maximum space between vents.

Solar Air Heater

Photo 4-4—Begin chipping away stucco until you find a mortar joint, then work in the perpendicular direction to find the other edge.

Photo 4-5—Once you have found two sides of a block, continue to chip away stucco until the entire block is exposed.

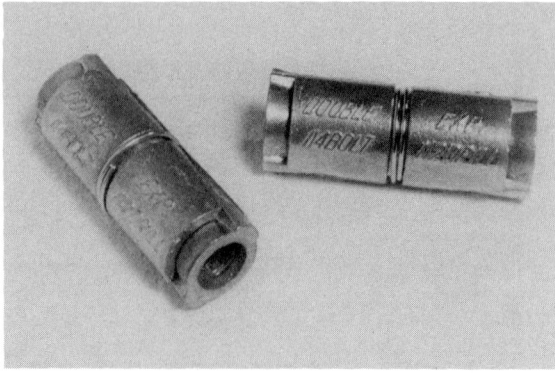

Photo 4-6—The best type of masonry fastener we have found looks like this. It uses a threaded bolt.

pencil in the exact location of the collector box when you have the first vent located. Be sure to leave at least two inches of clearance between the outer edges of the collector box and the vents.

With the masonry removed, drill a hole at each corner of each vent opening through the inside wall. Then go inside and remove the wall area covering the vents. Get the inside vent sides even with the opening in the masonry. If you have any furring strips or lathing strips in the vent opening, cut them off even with the opening.

First, remove the glazing frame and collector screen from the collector and mount the box temporarily, then take it down, work on it, and finally mount it permanently. We have found it best to mark the mounting holes on the back insulation first, being sure to leave room for the vents. Then drill shank holes through the back of the collector box at these positions. Before you mount the collector, cut away the back insulation at each bolt area, so the washer can tighten down against the plywood back.

Once you have the shank holes drilled and the insulation removed, put the collector in position against the wall, and mark one anchor bolt position on the wall. Be sure the collector is positioned exactly where you want it before marking the spot. Then take the collector down. This is a two-person job.

Photo 4-6 shows the anchor device we prefer. The bolt is 2 inches long and sized to fit the anchor. We have found that anchors requiring a ½-inch hole and ¼-inch bolts work best. We prefer these anchors as they use machine screws and are more exact than the lead shields or plastic liners often sold. They are worth looking around for. Be sure to use a masonry bit to drill into the masonry; it will give you a much more exact hole. With one anchor inserted in the wall, reposition the collector and fasten it to the wall with the one anchor you have installed. Then check the overall position of the collector on the wall, and, once you are sure it is where you want it, mark the position of the other anchor bolts on the wall. By fastening the collector on the wall with one another, you will be able to mark the other holes with a lot more accuracy. If you tried to mark all the anchor positions at one time, chances are the collector would shift slightly, and some of the holes would be off.

Before you take the collector down, go inside and mark the vent openings on the back of the collector. Then take the collector down and use a saber saw to cut out the vent openings in the back of the collector. After cutting the vent openings, you have to weather-strip the vent openings and the overall collector. We have found that a closed-cell foam weather stripping works best. Tape ½ inch thick and ¾ inch wide is ideal. Put this around the edge of the vents and on the back of the collector around the entire outside edge.

With the collector weather-stripped, and all the anchors inserted in the wall, you are ready to mount your collector to the wall permanently. Lift it into place, attach one anchor bolt, then position it and attach the remaining anchor bolts. Tighten the bolts well, but don't overtighten them. Be careful not to tear the weather stripping when positioning the collector.

With the collector up, caulk around the entire outside of the unit, taking special care to caulk the top edge well. This caulking will prevent any moisture from getting between the collector and the walls of the house. At this point you are finished mounting your collector. The rest of this chapter details wiring the collector, building the duct work, building the grilles, and installing the fans. Move on to finish your collector; the worst is behind you at this point.

DUCTS AND GRILLES

No matter what type of wall you have mounted your collector on, use this part of the chapter to finish your collector.

Blueprint sheet 5 details most of the information discussed in this part of the chapter. Use the illustrations on the blueprint sheet with the instructions to build your grilles and install the fans and back-draft damper.

Each duct is made out of four pieces of insulation board. The pieces should be cut to form four walls, each flush with the edge of the wall surface inside the house and the insulation surface in the collector box. With the pieces cut, position them to form a tight fit in each vent and tape the corners. Then tape the edges over the surface of the wall and the collector. Photo 4-7 shows the interior side of a finished duct. Extending the tape onto the wall surface helps to reduce infiltration and give the vent a finished appearance. Do the same on the collector surface and you'll end up with a duct that is firmly bonded to both the wall and the collector. The heat-activated tape we recommend is very easy to work with in this situation. It will stick to the painted surface of the collector as well as to the wall.

With the ductwork all taped, go outside and cut small patches of tape to cover the anchor bolts on the collector back. Tape should completely cover the areas where you previously cut away insulation for the bolts and washers.

With the taping complete, you are ready to install and wire the thermostat if you will be using your collector with fans. We recommend standard HPN heater cord, 16/2 for wiring, with a safety plug inside the house. Blueprint sheet 5 lists the materials we have

Photo 4-7—A taped duct should look like this. Use the insulation board for all sides of the duct, and tape all edges.

found to work best for the thermostat and fans.

Generically, the fans are known as computer fans, as they were originally developed for use with computers. The idea is to have 150 to 200 cubic feet of air pulled through the collector per minute. The thermostat will ensure that the collector does not get too cool and will turn the fans on only when heat is present in the collector.

Attach the thermostat to the thermostat-mounting block. Be careful not to split the block when attaching the thermostat.

Leave about 24 inches of wire loose at the thermostat and another 24 inches loose inside the bottom vent. Staple the remaining wire around the inside perimeter of the collector box with insulated staples.

Paint the vents and any patches you applied to the collector box with flat black paint.

Solar Air Heater

While the paint is drying you should weatherstrip the collector box. The best type of weather stripping to use between the collector box and the glazing frame is tubular vinyl gasket material. This material should be stapled around the edge of the collector box where the glazing frame will fasten down on it. Notch the weather stripping in the areas of the hanger bolts.

With the weather stripping in place, you are ready to attach the glazing frame permanently. First, replace the screen, then position the glazing frame over the hanger bolts and tighten it in place with the washers and cap nuts. That should complete your outside work. All you have to do is build the two grilles and install the fans.

The vent grilles we recommend are detailed on blueprint sheet 5 and shown in photo 4-8. The top grille should hold one or two fans; the bottom grille should hold the back-draft damper while restricting airflow as little as possible. The damper prevents cold air from falling back into the room when the collector is not producing heat. The fans are optional, but the back-draft damper is mandatory. Without the damper your unit would be a room heater during the day and a room cooler at night.

The type of grille frame we recommend is made from pieces of 1 × 3 material. We recommend a half-lap joint at each end, the same kind of joint you cut to make the glazing frame. There are a number of other ways to assemble the frames, but we recommend the method shown in illustration 4-6. It gives you a finished frame that is both strong and attractive.

Measure the openings of your vents, and build the grille frames so they overlap the vent openings by ½ inch on all sides. This overlap is an approximation. It can be more, but it should not be less.

With the two frames built, work on the top unit first, then the bottom one. If you will be using your collector without fans, simply build a second unit like the bottom unit, without back-draft damper, for the top.

The top unit needs a piece of ¼-inch plywood fastened to the back of the frame, with holes cut in it for mounting the fans. The plywood must be cut to fit exactly inside the vent opening and be fastened to the overlapping part of the frame.

Measure the vent to be sure two fans will fit before cutting the holes. Use the cutting template that comes with the fans to mark the holes, and cut them out, but don't install the fans until you have built the bottom grille frame and finished both frames.

The bottom grille frame is built the same

Photo 4-8—Top and bottom vent grilles. The top grille should house the fan or fans, while the bottom grille houses the back-draft damper. Hardware screen covers the bottom grille.

Illustration 4-6—The grille frame is made by joining four pieces of 1 × 3 material. Half-lap joints are used to fasten the frame together.

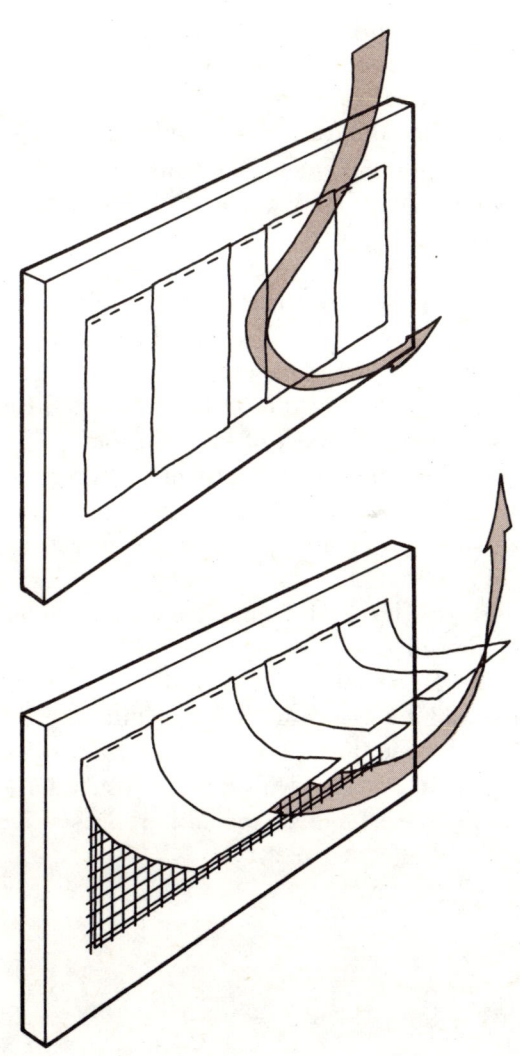

Illustration 4-7—The back-draft damper is a series of plastic flappers stapled to the back of the grille frame. The damper prevents cold air from entering the room when the collector is not in use.

as the top grille frame: four pieces of 1 × 3 with half-laps on the ends, overlapping the vent opening about ½ inch on all sides. With both frames built, paint or stain them to match your room's decor.

Next, fasten a piece of ½-inch hardware screen on the back of the bottom grille frame. The screen will protect the collector from foreign objects without hindering airflow very much. With the screen attached, you are ready to make the back-draft damper. This is nothing more than a series of plastic flappers that are pulled up and out of the way by air flowing into and up the collector and are forced against the screen by cold air falling out of the collector, preventing this air from getting into the room. The plastic flappers should be cut the same length as the vent opening and about 3 inches wide.

The flappers should be attached in two layers. The first layer should go from one side to the other, with about ½ inch of space between each flapper. The flappers in the second layer are centered over the spaces in the first layer. The finished effect should be two layers of overlapping flappers, as shown in illustration 4-7. Take your time building the back-draft damper, and it will keep your home good and warm nights. The best material to use for the flappers is a standard 2-mil household garbage bag. One bag will make many more flappers than you will need.

Now turn your attention to the top grille. Mount the fans on the plywood, according to the template provided with the fans. Then connect the wiring, and attach the grille to the wall. Blueprint sheet 5 shows how the fans should be wired. One wire should go directly to the fans, the other to the thermostat. From the thermostat, the wire should then go to the two fans. This will enable the thermostat to turn the fans off when heat is not being produced. With the wiring done, attach the grille to the wall, and the top of the collector will be done.

Before installing the bottom grille, pull the wire out of the vent through the screen at a bottom corner of the grille, and connect a plug to it. Be sure the wire doesn't interfere with the back-draft damper operation. If you know how to wire, you may want to snake the wire down through the wall and wire it directly. Then attach the bottom grille to the vent, being careful not to damage the wire and plug. Plug in the unit during the months of the year when you want heat, and the thermostat will turn on the fans whenever heat is being produced.

That completes your solar hot air heater. Congratulations—we hope you enjoy it, and your free heat, for many years to come.

Section III Blueprints

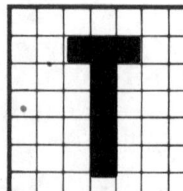

This section is the heart and soul of the book. The six blueprint sheets bound in after this page should answer all questions about building a solar hot air heater. Use the blueprints hand in hand with the instructions when building your collector. The two work as a team; keep them together.

Each sheet is folded and perforated. Crease each sheet along the perforations, and tear them out one at a time. There is also an envelope bound into the book; don't tear it out. The envelope is there to give you a place to store your blueprints with the book when you're finished. You'll most likely want to come back and use the book to build another heater in later years, and this way you'll always have your blueprints handy. The envelope serves to protect the blueprints, yet allows you to take them with you to the shop when you build your heater.

There are six blueprint sheets with Rodale's *Solar Air Heater*. Most likely you will not use all the sheets, unless you build both a vertical and a horitzontal heater. Once you have decided which style heater to build, use only those blueprint sheets, to avoid getting confused.

Blueprint Sheet 1—Vertical Collector Box. This sheet details all the pieces for a vertical collector box. Use this sheet with the first part of chapter 2. Pay close attention to the instructions; you'll be cutting some pieces on blueprint sheet 3 at this time.

Blueprint Sheet 2—Horizontal Collector Box. This sheet details all the pieces for a horizontal collector box. Use this sheet with the second part of chapter 2. Pay close attention to the instructions; you'll be cutting some pieces on blueprint sheet 4 at this time.

Blueprint Sheet 3—Vertical Glazing Frame. This sheet details the pieces for a glazing frame for a vertical heater, including the collector screens and the glazing materials. Use this sheet with the first part of chapter 3 and blueprint sheet 1. Some measurements are only approximate. It is best to measure the actual opening and cut to fit.

Blueprint Sheet 4—Horizontal Glazing Frame. This sheet details the pieces for a glazing frame for a horizontal heater, including the collector screens and the glazing materials. Use this sheet with the second part of chapter 3 and blueprint sheet 2. Some measurements are only approximate. It is best to measure the actual opening and cut to fit.

Blueprint Sheet 5—Vent Grilles. This sheet shows a typical vent grille for either a horizontal or vertical heater. These are examples of only one type of installation. Follow the instructions in chapter 4 to custom-install your individual heater on your home. The measurements of the pieces will vary according to your installation.

Blueprint Sheet 6—Cutting Diagrams. This sheet has separate cutting diagrams for vertical and horizontal collectors. The shaded pieces are for the instructions in chapter 3, while the unshaded pieces are for chapter 2. Each cutting diagram has a lumber list to use when shopping. Consult the materials section for advice on what types of wood to buy for your heater.

This completes your *Solar Air Heater* plans book. We are sure that the designs presented here will work. You may be able to use some recycled materials to reduce your cost. We have presented the best way to build the heater, using the best of materials and placing a premium on the quality of the design of the heater. Your collector should last many years and provide many years of free heat; build it well.

If you have any questions on the construction of the heaters, contact the Reader Service Department, Rodale Plans, Rodale Press, 33 East Minor Street, Emmaus, PA 18049. Please don't ask us about the availability of materials; we've given you all the addresses we have in the Materials section and on blueprint sheet 5.